Grow Food Anywhere

Grow Food Anywhere

How to plant the right crops in the right place and help your garden thrive

Lucy Chamberlain

Contents

Chapter 1
What this book is about   6

Chapter 2
Learn to read your plot   14

Chapter 3
The importance of soil   50

Chapter 4
Matching crops to zones   58

ZONE 1 Sunny and sheltered   60
Peaches, nectarines, and apricots; sweet cherries; basil; runner beans; strawberries; tomatoes; Malabar spinach; peppers; aubergines; sweet potato

Ten other star performers   72

Project: Fan training fruit in the sun   74

ZONE 2 Sunny, open, and dry   76
Asparagus; salsify and scorzonera; summer purslane; lima beans; figs; agretti; goosefoot; grapes; globe artichoke; amaranth/quinoa

Ten other star performers   88

Project: Plant supports for windy sites   90

ZONE 3 Sunny and moist 92
Winter squashes and pumpkins; French and drying beans; Asian broccolis; courgettes and summer squashes; broccoli; potatoes; lettuces; dahlias; pears; Florence fennel

Ten other star performers 104

Project: Waving goodbye to weedy plots 106

ZONE 4 Open and cold 108
Onions and shallots; carrots and parsnips; broad beans; plums, damsons, bullaces, and sloes; beetroot; winter kales and cabbages; blueberries; leeks; turnips, swede, and kohlrabi; apples

Ten other star performers 120

Project: Planting an edible windbreak 122

ZONE 5 Part shade 124
Salad rocket; annual spinach; raspberries; elder; mint; Chilean guava; Swiss chard and perpetual spinach; Texsel greens; pak choi and relatives; Szechuan pepper

Ten other star performers 136

Project: Creating a beautiful wild berry patch 138

ZONE 6 Shady and wet 140
Celeriac and celery; rhubarb; winter purslane; blackcurrants; prickly heath; parsley; hosta; dog's tooth violet; mizuna and mibuna; minutina

Ten other star performers 152

Project: Growing your own mushroom "loggery" 154

ZONE 7 Shady and dry 156
Gooseberries; red, pink, and whitecurrants; alpine strawberries; horseradish; hazelnuts; Japanese wineberry; garlic mustard; patience dock; perennial nettle; bamboo

Ten other star performers 168

Project: Making a miracle mulch for healthy roots 170

ZONE 8 Indoors 172
Dwarf tomatoes; pea shoots; microleaves; prickly pear; lemongrass; citrus; ginger and relatives; pomegranate; agave; dwarf chilli peppers

Ten other star performers 184

Project: Creating an artificial light growhouse 186

Chapter 5
Tweaks to optimize your plot 188

Chapter 6
Troubleshooter 204

Glossary 214
Useful resources 216
Index 217
Acknowledgements 223

What this book is about

//
Give crops what they need

Just as buddleias enjoy the sun and wood anemones revel in shade, edible plants evolved in particular planting sites. Consciously thinking about these when we plant up our vegetable gardens and allotments will guarantee robust, content crops that will reward us with the very best harvests – no more disappointments.

Our current mindset for growing our own food is dominated by planting what we love – after all, there isn't much point in growing edibles that you won't eat. But a crop we love to eat may not love where we plant it in the garden. Dare I ask, how often is this even a conscious choice? Most of us are guilty of plonking a vegetable plot where we have room for it, often away from the house as an afterthought, playing second fiddle to our flowerbeds and social areas. The consequences could be sulking crops shoehorned into places they don't enjoy. That's why there is a real need to assess and fine-tune our attitude towards food growing.

Are we thinking of edibles in the wrong way?

As gardeners, we are forever taught that clematis thrive with shaded roots, rhododendrons prefer acidic soil, and pelargoniums revel in full sun, so perhaps we should be more mindful that edibles have specific growing needs, too. If we start giving them the conditions they enjoy, they will be naturally vigorous, robust, and healthy, ready to deliver maximum yields, and surely that's the objective when growing your own food.

Crops thrive in specific locations

Edibles evolved in specific climates, just like ornamental plants. For example, figs revel in Mediterranean sunshine, redcurrants are happy in the shade of Western Europe, and celery appreciates marshland soils of the Mediterranean basin. Sea beet (the parent of beetroot and Swiss chard) is found on sandy European shorelines, whereas wasabi originates from the permanently damp riverbeds of Eastern Asia. These adaptations offer brilliant opportunities for gardeners, and understanding a plant's natural growing conditions will help us choose crops that will thrive in any given part of our garden, delivering fantastic yields as a result. We really can grow food anywhere!

Understanding individual crops' needs allows you to focus your efforts on what they enjoy. Earmark impoverished sites for robust crops that don't need mollycoddling. Simultaneously, be more targeted with resources by reserving moist soil for thirsty crops, and fertile beds for hungry feeders.

Work with nature

Positioning your edibles where they will thrive will reward you with healthier plants and bigger harvests.

The concept of giving crops what they need isn't a one-sided arrangement – we gardeners benefit hugely! You'll actually need to do less work – less watering, less feeding – working with nature takes the effort out of growing your own food, and the positive results will speak for themselves.

Conserve your resources
For example, position thirsty crops like bulb fennel or hostas away from a dry soil – otherwise you'll be forever throwing water at them, they will struggle, you'll get disheartened, and you'll also be wasting that resource. Instead, reserve dry zones for drought-resistant crops – there are plenty to choose from. Asparagus, orache, seakale, salsify, tree spinach, and lima beans are just a few examples. Equally, a crop that prefers sharp drainage and good air movement, globe artichoke, agretti, or horn of plenty, for example, could rot in a shady, damp spot – that soil moisture would be far more beneficial to another, more appropriate edible such as celeriac or sorrel.

Protect the environment
Economize on fertilizer, reserving any animal manure or boxed feeds for hungry squashes and pumpkins, rather than wasting it on seakale or salsify, which are deep rooted and can hold their own on poorer soils. Adding fertilizers unnecessarily is not only detrimental to your budget, unused nutrients can also leach into groundwater and harm the wider environment.

Gain better harvests
Working with nature, by definition, gives better yields. Tomatoes and aubergines would offer scant harvests in cool shade, yet position them against a south- or west-facing sunny fence and they'll produce an almighty, delicious glut of fruits. Annual spinach, conversely, would quickly run to seed in such blazing summer heat but give it a cool, shady border and it will reward you with a prolonged and abundant lushness of foliage.

   The bottom line? Understand your crops and they will be healthier and give you larger yields. Ultimately, this more focused, holistic mindset benefits everything – and everyone.

Grow an amazing range of edibles

Your garden contains many individual microclimates, each one able to provide different crops with exactly what they need – how varied and fantastic is that? Embrace these elements, and your home-grown culinary repertoire will become far more diverse and exciting – you'll also be introduced to delicious crops that you never knew existed.

Shady fences, beautifully sheltered suntraps, dry beds under trees, permanently damp corners, wind-beaten alleyways – most gardens contain a range of growing environments, or "zones", and there are edibles that will thrive in them all. You might not have come across many of them, so within this book let's set up some life-long introductions.

New "must-grow" crops

Say hello to ground-hugging summer purslane and New Zealand spinach that both adore the most arid and windswept of "unpromising" beds. Or why not get acquainted with shuttlecock ferns, Caucasian spinach and Solomon's seal that would love nothing more than to colonize that dank, dark corner behind your shed.

As you learn to identify the different growing zones in your garden, you'll realize that every plot offers an array of opportunities for growing a huge selection of exciting and, above all, delicious edibles – each year, I grow over 150 different edibles on my 10 × 10m (33 × 33ft) plot. Please, don't label a single area in your garden as troublesome again – there's an edible that will at the very least tolerate it, and at best love it.

Move towards self-sufficiency

As you become familiar with the diverse range of edibles that can thrive in your garden you'll notice something: the gluts and dearths associated with growing your own food become less prominent. Cultivating smaller numbers of a wider range of crops smooths out the supply of food to the kitchen. Thank goodness for that! The late summer dramas of drowning in and having to offload courgettes and runner beans will be replaced by days celebrating the first pickings of agretti, Malabar spinach, and Japanese wineberries. Friends, neighbours, and relatives will heave a collective sigh of relief.

Clockwise from top left: feijoa, sea buckthorn, perilla, and Swiss chard will all bring a beautiful diversity of edibles to your plot, each thriving in its own preferred planting zone.

Learn to read your plot

To take the first steps towards growing food anywhere, start by getting to know the microclimates in your garden. That shady, disused corner is perfect for growing lush, velvety spinach. The wind-battered front balcony? Agretti and purslanes will thrive there. And the dumping ground for garden waste – courgettes and squashes will love sinking their teeth into your compost heap.

Discover your garden's microclimates

A crop for any spot

Every garden, allotment, community space, and balcony contains areas that are perfectly suited to certain crops. Square or rectangular plots, large or small, can possess every microclimate you need – sun, shade, wet, dry – it's all there for the taking. Essentially, these spaces are smaller versions of more traditional kitchen gardens.

Compact plots such as balconies, patios, and courtyards offer great growing environments in miniature. Clothe the fantastic wall space in grapevines where the sun beams through, or Japanese wineberries and redcurrants where it doesn't. Shade-loving hostas, or sunbathing tomatillos, can be easily nurtured in pots, and what basil wouldn't love to live in a sunny window box alongside thyme, sage, and oregano?

Mapping your plot

No two plots are the same, so this chapter focuses on a variety of garden environments to arm you with excellent mapping skills you can apply to your own growing space. By "reading" your plot's landscape you'll feel more connected to it, and more confident. You'll also be perfectly placed to choose crops that will thrive in it. See Chapter 4 for my suggestions of a wide range of wonderful crops to grow in eight example microclimates, or "Zones".

The objective when mapping your plot is to gain an understanding of where the sunny south- and west-facing corners are. These suntraps are perfect for cultivating sugar-packed apricots, figs, and peaches. Chillies, lemongrass, and tomatoes will positively beam at such a naturally warm microclimate – no greenhouses, polytunnels, or propagators required. However, north- and east-facing shadier spots are no longer "the enemy". Alpine strawberries, rhubarb, sorrel, gooseberries, land cress, and chervil will all thank you enormously for the cool relief, come summer.

Tomatoes enjoy sun and shelter (above left); grapes bathe in sun and sharp drainage (above centre); squashes swell brilliantly with sun and ample moisture (above right); onions will fend well in open cold sites (right); Swiss chard produces lush foliage in part shade (far right); gaultheria enjoys moist shade (below left); bamboo will yield in drier shade (below centre); agaves appreciate a frost-free spot during winter (below right).

A sunny spot in the garden is like the horticultural equivalent of winning the lottery. Light is one of the key components for photosynthesis, so having it in abundance is a gift. Just add air and water and you have the perfect trio for the harvests to roll in.

Sun

Rapid growth and maximum sweetness

Fruits that bathe in sunlight such as grapes (right) and peaches (below) respond by producing harvests rich in sugars. Vegetables such as tomatoes (bottom right) and butternut squashes (top right) can contain 10 per cent sugar, and dahlias (centre right) yield tubers that mellow in storage.

A plant's growth rate is largely determined by light levels. Although some species have evolved to grow at low intensities, most edibles – especially those that flower and fruit – will respond positively to being bathed in sunshine.

Maximizing flavour

Fruiting crops develop their highest sugar levels in prolonged full sun, and this will directly improve their flavour. Training peaches, apricots, figs, and nectarines along a sunny wall will allow them to develop that rich, deliciously honeyed taste. Most sweet-tasting fruits contain 8–10 per cent sugar, but dessert grapes and figs can manifest as much as 16 per cent if they receive sufficient levels of sunshine.

Even sweet vegetables, like tomatoes and butternut squashes, can contain anywhere from 3–10 per cent sugar, so placing them in a sunny spot is a good way to achieve the best flavour.

LEARN TO READ YOUR PLOT

Storing crops correctly to optimize sweetness

Some crops carry on ripening once they have been picked, mellowing and sweetening in storage, which boosts their sugar levels when growing in temperate climates. Ripen peaches on a sunny windowsill to raise sugar levels from 10–13 per cent. (Keeping them inside also protects them from wasps as sugar levels peak.) Tomatoes, too, can also get this treatment – I've recorded their sugar levels rising from 8–10 per cent using a Brix refractometer (see page 214). Sweet potatoes, late-season pears and apples, plus winter squashes, all accumulate sugars during long-term storage somewhere cool, frost-free, and dark.

Other crops, such as traditional sweetcorn and pea varieties, can lose sugar content as soon as they're harvested, quickly converting it to starch. In these cases, freeze crops as soon as possible after picking to conserve their sugar levels.

Improving plant hardiness

High levels of sunlight also mature and ripen the wood of fruit trees, canes, vines, and bushes. This increases winter hardiness, preventing dieback, and encourages more fruit buds to form the following year. Food stores of bulky root crops that develop sweet-tasting corms, bulbs, or tubers (sweet potatoes, yams, and yacon, for example) are also maximized via a position in full sun.

If there's one thing I want to get across in this book, it's that shade isn't a disadvantage. Shade-loving edibles can be described as "lush", "verdant", "abundant", "soft", and "palatable". So let's celebrate these unappreciated nooks.

Shade

Everlasting abundance and cool relief

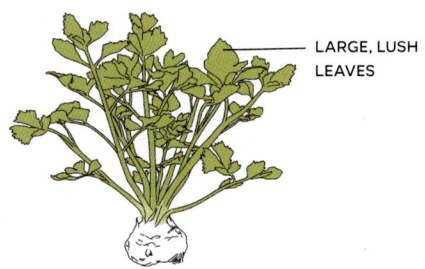

LARGE, LUSH LEAVES

Many flavoursome edibles thrive in shade, such as celeriac (right), alpine strawberries (centre right below), bamboo shoots (centre right above), gooseberries (far right), and redcurrants (below right). Beds of leafy crops (below) perform well in the coolness of a shady spot, especially in summer.

Shadier plots can embrace their natural ability to encourage generous succulent foliage (left).

Certain crops would scorch or become leathery in full sun. Choose edibles that have evolved in shaded habitats such as woodlands, the base of cliffs, and the bottom of ravines, like alpine strawberries, shuttlecock fern, or hosta, and harness their natural adaptations.

Shade-loving crops frequently have large leaves and canopies that capture as much light as possible. Shaded environments are often unhindered by buffeting winds, enabling leaves to grow large.

Dappled, part, or deep?

Hedgerows and woodland margins provide light or dappled shade, where plants receive sun for part of the day – shrubberies provide a similar habitat. Partly shaded conditions, in a deciduous forest or under your sizeable cherry tree, provide

LEARN TO READ YOUR PLOT   21

more or less constant shade throughout the day. Deeply shaded areas, such as a deep ravine or behind your garage, never receive sunlight.

Shade can be a blessing during heatwaves. Avoiding prolonged or extreme UV rays is necessary for many crops, unless their leaves are covered in protective oils, waxes, or hairs.

Cool and moist

Shade frequently offers cool, moist soil – excellent for slow-to-bulk-up and thirsty edibles such as celery, celeriac, and parsley. Just as ornamental clematis like having their feet in the shade and their heads in the sun, so do raspberries, honeyberries, loganberries, blackberries, and mashua.

Tarter fruits can accommodate a shadier location. Reserve your sunny borders for sugary figs and grapes, the likes of sour rhubarb, currants, and gooseberries are very happy growing in low light. Think, too, of the love for foraging for berries in Scandinavian forests – lingonberries, crowberries, and bilberries all thrive in woodlands. If dry shade is your concern, jostaberries, wineberries, and worcesterberries are just a few fruits that will yield, and potherbs, such as garlic mustard, nettle, and sweet cicely can join them.

How to map the sun and shade

Sunlight is one of the key requirements for photosynthesis, and the warmth that it brings encourages optimum growth rate and yields. However, some edibles find excessive levels of heat and sunshine difficult to cope with, so understanding where this falls in your garden is key to gaining good harvests.

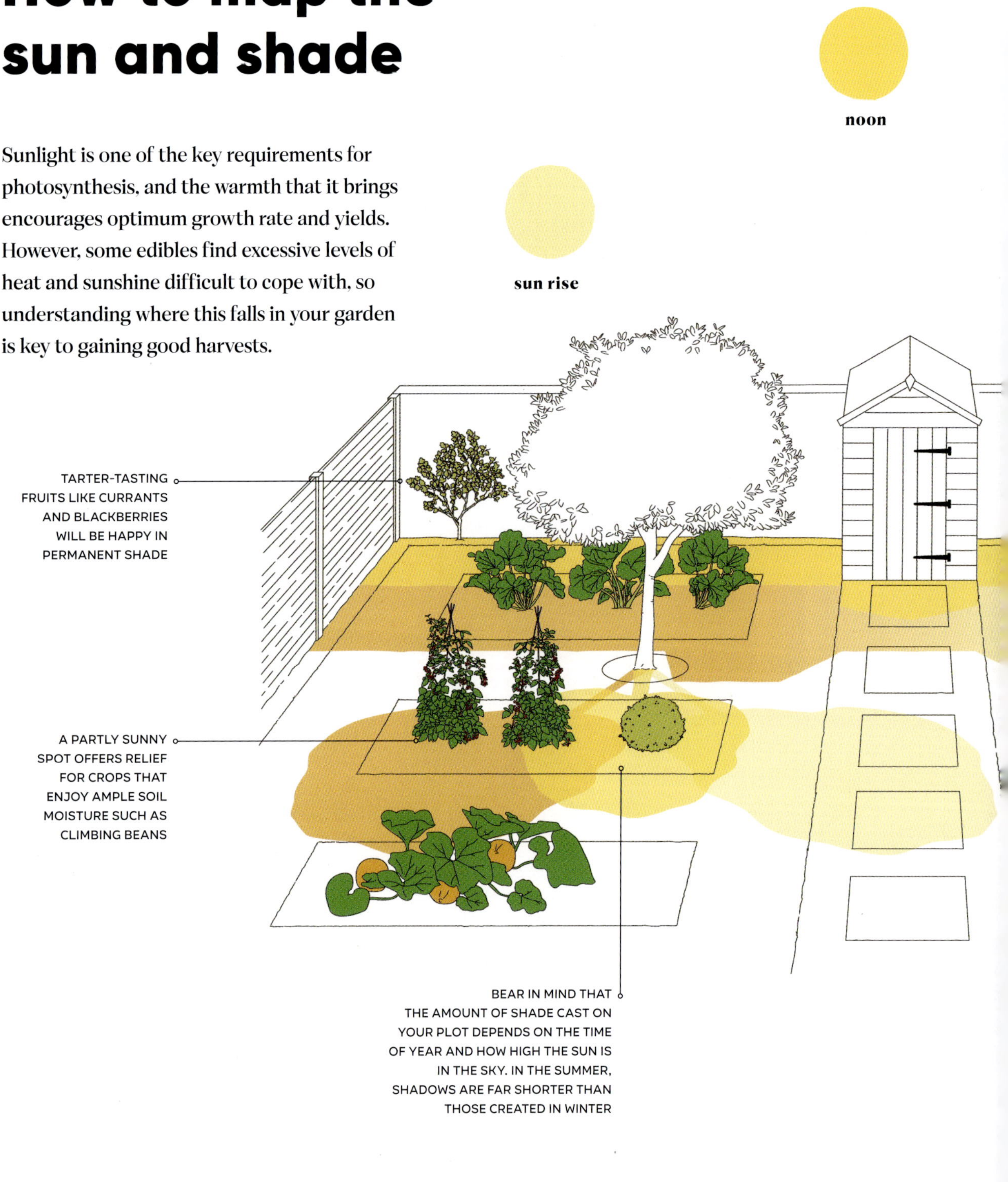

noon

sun rise

TARTER-TASTING FRUITS LIKE CURRANTS AND BLACKBERRIES WILL BE HAPPY IN PERMANENT SHADE

A PARTLY SUNNY SPOT OFFERS RELIEF FOR CROPS THAT ENJOY AMPLE SOIL MOISTURE SUCH AS CLIMBING BEANS

BEAR IN MIND THAT THE AMOUNT OF SHADE CAST ON YOUR PLOT DEPENDS ON THE TIME OF YEAR AND HOW HIGH THE SUN IS IN THE SKY. IN THE SUMMER, SHADOWS ARE FAR SHORTER THAN THOSE CREATED IN WINTER

LEARN TO READ YOUR PLOT

Start with the fundamentals when mapping your plot. To get your bearings, use a compass to establish which direction is north, then observe where the sun rises (in the east), and where it sets (in the west). North- and east-facing aspects will mostly be shady, whereas those looking towards the south and west have potential to be far sunnier. Map where the sunshine and shade fall on your plot at various times of the day. Taking photos is a simple way to discover this. Note on your map which areas are in permanent shade and full sun – these are your extreme microclimates. Dappled or temporary shade – that offered by deciduous trees or low walls and hedges – creates more modest growing environments. Take gradients into consideration, too. Plots sloping openly towards the sun will bask in more sunlight than level gardens or those tilting towards the north.

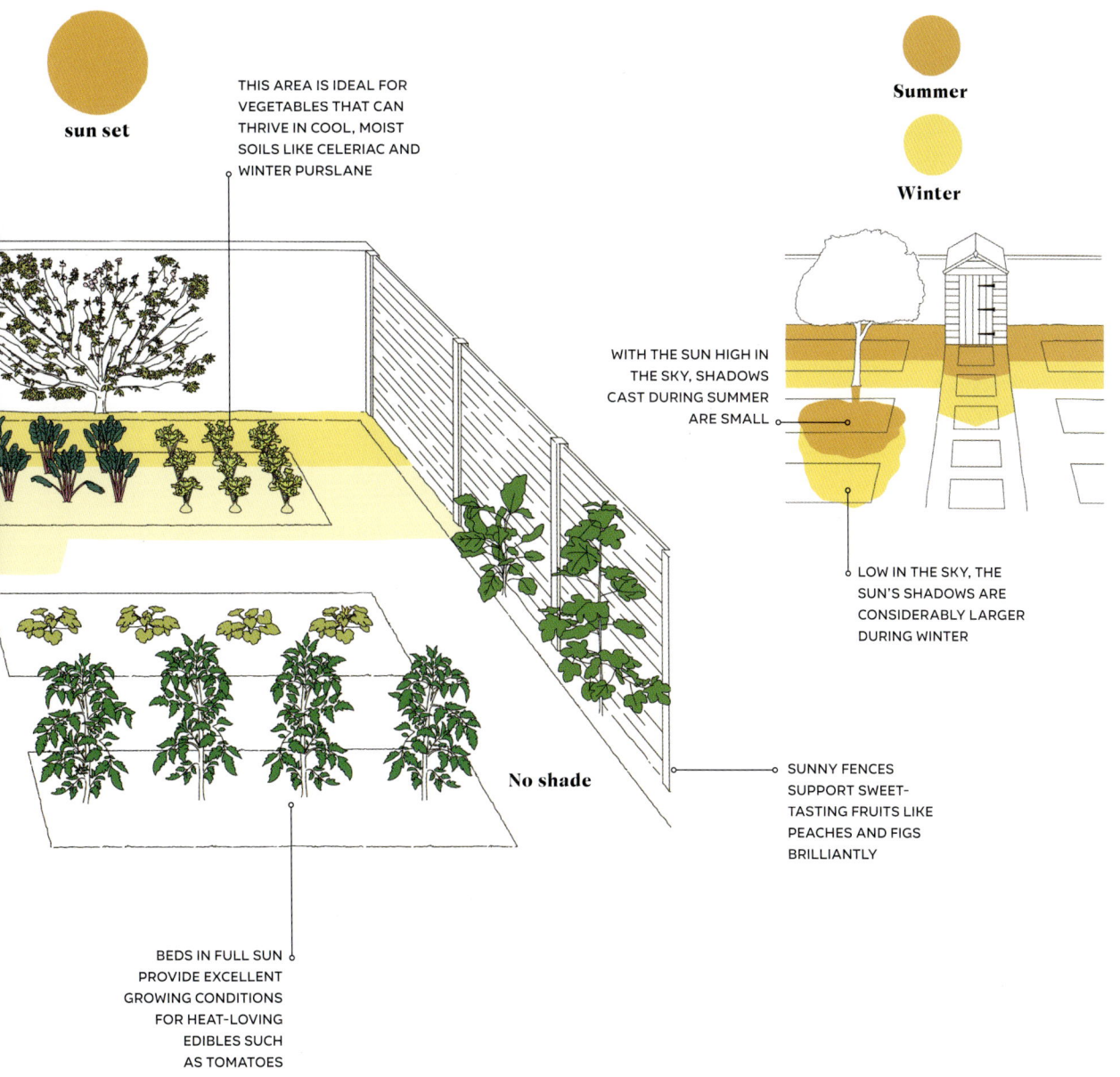

A moisture-retentive plot is brilliant for growing edibles that have adapted to wetter conditions and it frequently encourages gargantuan yields.

Wet

Steady growth and amazing yields

Bulky crops such as winter squashes (right) and courgettes (far right) contain 95 per cent water. Gaultheria (below) will give abundant berries on damper soils, and parsley (below right) will produce generous foliage.

It's staggering, but the harvestable parts of some vegetables are almost entirely – 95 per cent – made up of water. Not surprisingly this includes courgettes, cucumbers, and pumpkins. Add warmth, and you can see how glorious summertime gluts of cucurbits can become commonplace. Crops like watercress and wasabi that have evolved along streambeds, banks, and riversides enjoy free-flowing water around their roots. This moisture can be permanent or seasonal. Think, too, of the stagnant conditions of bogs and marshland. Blueberries, cranberries, lingonberries, and gaultheria all originated from wetter soils – so you see how wet environments can be home to a variety of crops.

Luscious leafy veg

Abundant foliage is easily attained when ample water is available. Leafy salads and herbs that

bulk up quickly like mizuna, minutina (buck's-horn plantain), parsley, and American land cress will reward you hugely (and quickly, in warmer weather) if positioned in a damper location. Annual spinach and rocket are less keen to run to seed or "bolt", if their roots are protected from drought stress.

Fertile boglands

Bogland fruits like cranberries and blueberries are ideal for low-lying areas where moisture is more likely permanent. Winter purslane, chervil, Florence fennel, mitsuba and mooli are all happy in moist – but not stagnant – soil. Dig bulky organic matter such as composted bark into the bed to achieve this. If your plot is on a gradient, create a soakaway (a gravel-filled pit) at its lowest point to wick excess moisture away from other areas. Raised beds also offer improved drainage. The slightly elevated roots can access the water reserves below them, yet remain aerated and healthy.

Some water is essential for plant life, but certain edibles evolved to survive with very little. Plants found in dune, scrub, desert, and savannah zones thrive in low rainfall conditions with excellent, sharp drainage – not a rotted tuber or crown in sight.

Dry

Extreme resilience and tender success

NARROW NEEDLES

WAXY SKINS

SUCCULENT LEAVES

Various edibles have leaf adaptations to boost their drought resistance.

Drought-resistant plants have physical adaptations – leaf spines, hairs, and glaucous, oily, or waxy skins that safeguard against desiccation. Narrow, needle-like leaves lose less water in hot, dry weather and water-packed succulent leaves and roots act as reservoirs. Robust, rot-proof growth is commonplace in dry climates.

Resilience during drought

If climate forecasts are realized, drought-tolerant edibles will be advantageous to gardeners. The UK summer of 2022 saw many areas devoid of rainfall for 6 months. Some edibles failed but others thrived. Established grapes, figs, amaranth, tree spinach, globe artichokes, and purslane all performed brilliantly in my garden. Direct-sown annual crops show more tolerance to drought than transplanted ones.

Dry gardens are excellent for overwintering borderline hardy crops – pomegranate, feijoa, and loquat for example – or those that hate winter waterlogging, such as strawberries and broad beans. Perennial edibles with fleshy roots or tubers such as scorzonera and Jerusalem artichokes, pass through winters unscathed given adequate drainage.

Crops will quickly recover from summer flash flooding if drainage is sharp and the climate predominantly dry. Drought-proof adaptations include, for many succulents and cacti, the ability to take up and store lots of moisture when rare rains fall.

The value of mulch

Leaf litter, retained dead foliage, and even rocks can act as mulch, conserving moisture where it is needed – around the rootball. With drier soils, having a source of organic matter in your garden, via a compost heap or leaf mould pile, pays dividends for crops wanting to spread their roots in search of water. Conversely, a dry mulch of stone or gravel, locks in moisture for edibles that evolved in low fertility soils. Either way, aim for a layer at least 4cm (1½in) thick so that the roots benefit sufficiently.

LEARN TO READ YOUR PLOT 27

Clockwise from top left, summer purslane, amaranth, asparagus, agretti, and salsify are all adapted to thrive in drier soils.

How to map wet and dry areas

Adequate moisture is essential for plant growth, especially for bulking up crop yields. While permanently wet plots prove fatal for most plants, some edibles are happy in boggy soil. Other crops excel in drought conditions, where resulting growth is robust, resilient, and surprisingly succulent.

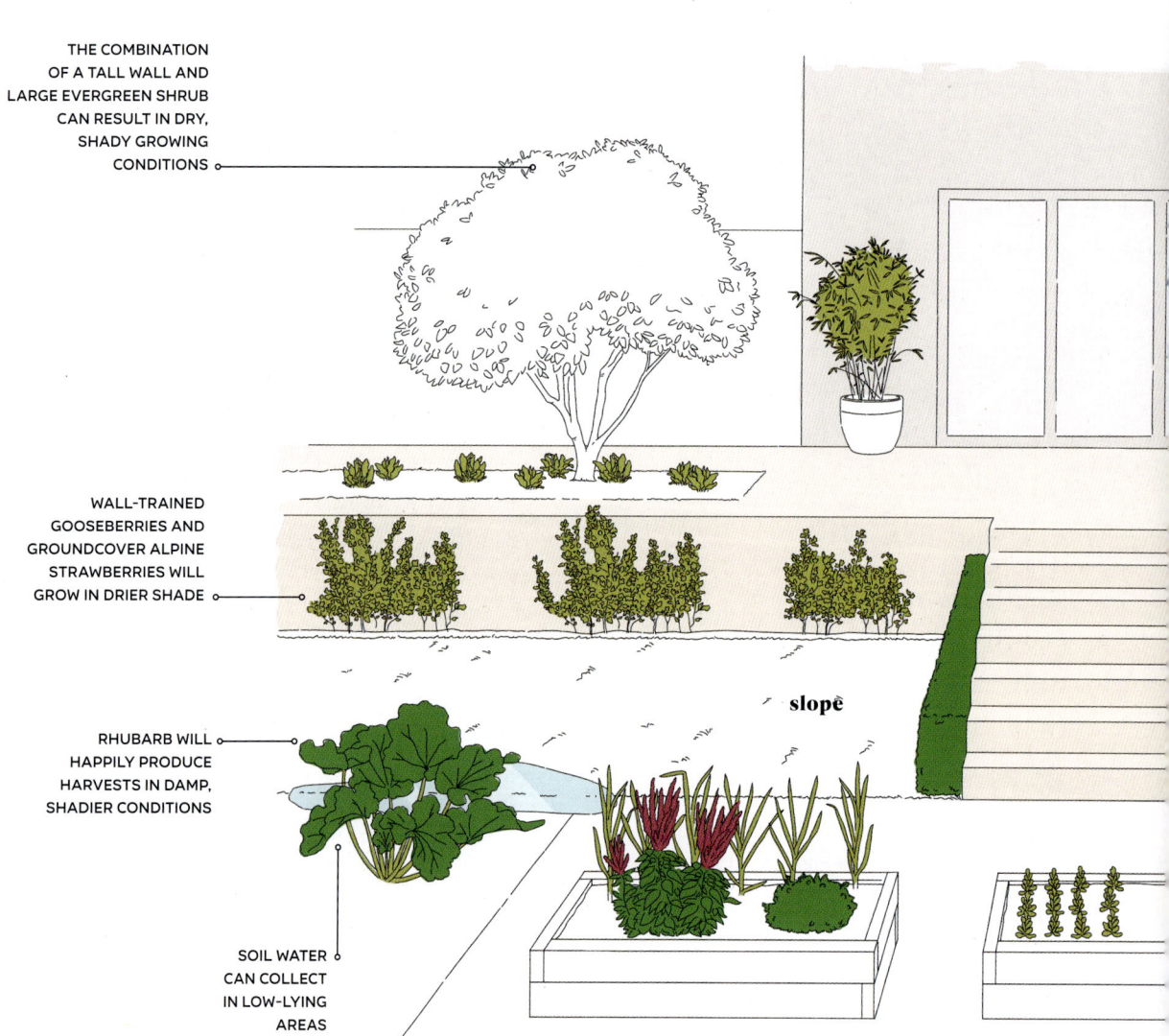

THE COMBINATION OF A TALL WALL AND LARGE EVERGREEN SHRUB CAN RESULT IN DRY, SHADY GROWING CONDITIONS

WALL-TRAINED GOOSEBERRIES AND GROUNDCOVER ALPINE STRAWBERRIES WILL GROW IN DRIER SHADE

RHUBARB WILL HAPPILY PRODUCE HARVESTS IN DAMP, SHADIER CONDITIONS

SOIL WATER CAN COLLECT IN LOW-LYING AREAS

slope

LEARN TO READ YOUR PLOT

Soil moisture levels can either be constant, or seasonal. Start by looking at any gradient on your site – are there any low-lying areas? These may be permanently moist. Digging a small 30cm (12 in) wide and deep excavation pit can help determine continually waterlogged areas. Equally, gardens at altitude are likely to drain well and may have dry soil, unless fed by underground springs or regular rainfall. Do puddles form in certain zones after heavy rain? Note temporary conditions like this on your map. Areas exposed to fierce summer heat or desiccating winds can be earmarked for drought-proof edibles. Raised beds, too, can often be dry due to their elevated levels. Large deciduous trees such as oak and willow, will readily drink moisture from the soil during summer, but remove none during winter when they are dormant. Evergreens like conifers, yew, and box can permanently dry out soil due to their mat-like system of continually thirsty roots. Soil type (see page 52) can also have a notable effect on soil moisture levels.

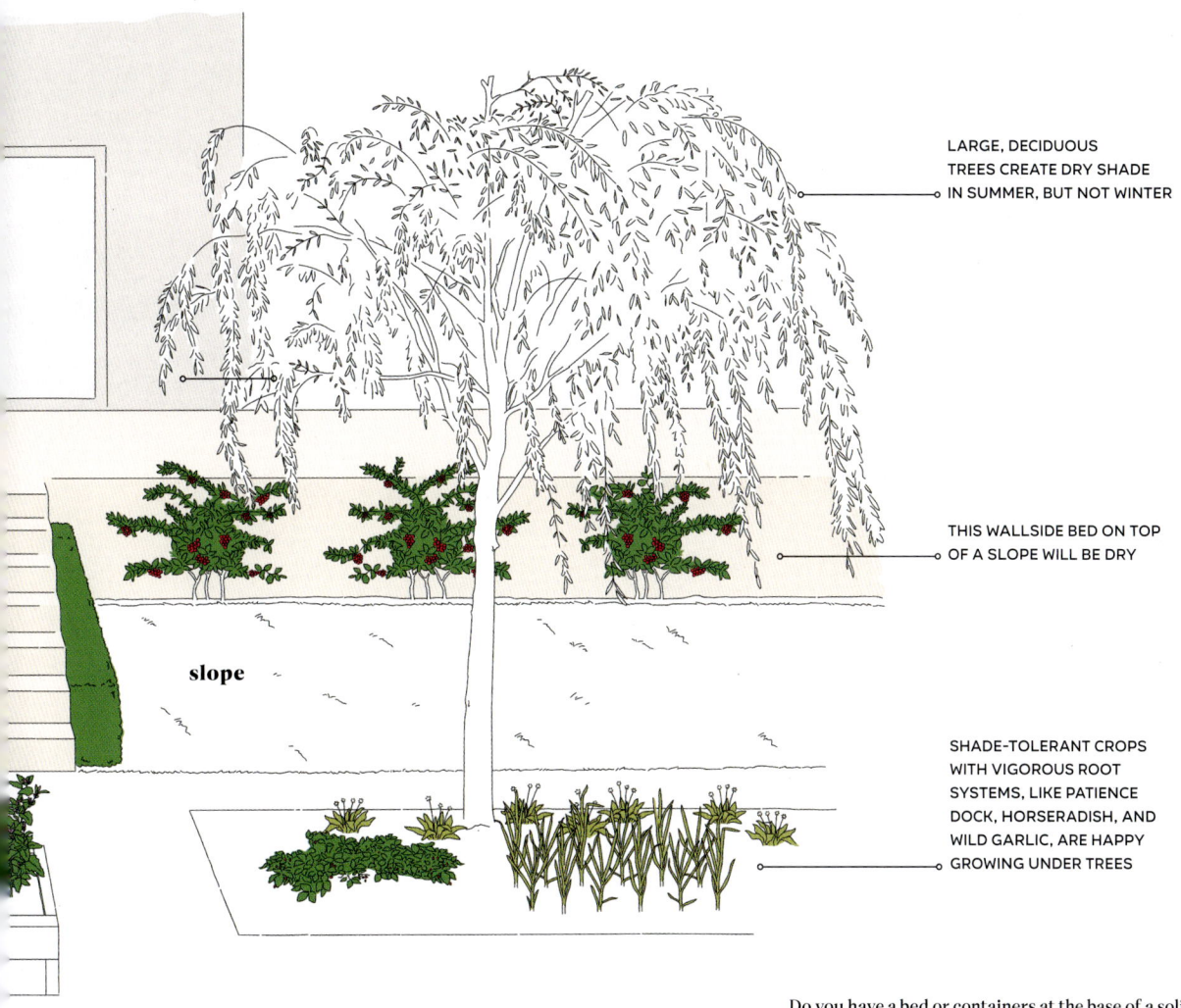

LARGE, DECIDUOUS TREES CREATE DRY SHADE IN SUMMER, BUT NOT WINTER

THIS WALLSIDE BED ON TOP OF A SLOPE WILL BE DRY

SHADE-TOLERANT CROPS WITH VIGOROUS ROOT SYSTEMS, LIKE PATIENCE DOCK, HORSERADISH, AND WILD GARLIC, ARE HAPPY GROWING UNDER TREES

Do you have a bed or containers at the base of a solid wall, garden building, or fence? "Rain shadows", where the soil becomes very dry, can form in such areas.

Cool conditions bring welcome relief in heatwave summers, and don't presume that temperatures below a pleasant 17–19°C (63–66°F) are adverse for your garden.

Cold

Brilliant blooms and supreme hardiness

Prolonged periods of sub-freezing temperatures limit what perennial edibles you grow. However, many crops from temperate, subarctic, or even arctic environments rely on cold temperatures to trigger key processes. Flowering of many fruiting crops, apples for example, is synchronized and boosted only once plants have been exposed to sufficient cold – this process is known as "vernalization".

Success in summer

Cool summer conditions alleviate heat stress in crops from temperate climes. Bolting in spinach, rocket, lettuce, and other salads is a frustrating summer phenomenon occurring just when you want to eat them. Shade (and the accompanying drop in temperature) provides welcome relief for these crops.

Severe winters allow us to identify areas that hold onto cold. Shadier zones can remain frozen during cold snaps as they are slow to warm without the sun's rays. Choosing especially hardy plants for these locations will prove invaluable.

Crops that have evolved in cooler climates will naturally thrive in temperate gardens. Clockwise from far right: apples, blueberries, kale, beetroot, and leeks are all sensible choices for more exposed plots.

Choose hardy plants

We treat crops from tropical climates with care in temperate regions, and take precautions to overwinter them using insulating materials. But many familiar edibles are extremely hardy, so working with these means less effort, and less disappointment. Hardy plants evolved certain physiological responses to cold temperatures, which are triggered during the onset of autumn and, ultimately, winter (we set off a milder version of this reaction when we "harden" plants off from greenhouses to life outside). Soluble sugars glucose and fructose accumulate in cells to reduce the movement of liquids and reinforce membrane stability – both actions protect cells against frost damage. It's often cited that Brussels sprouts and parsnips taste sweeter after a frost, and this is why.

On colder plots, choose edibles that evolved in cold climates. Rhubarb from Scandinavia, raspberries from the Arctic, and blueberries from North America will all shrug off chills. Hardy brassicas such as kale, Brussels sprouts, and Savoy cabbage, alongside broad beans, beetroot, leeks, carrots, and parsnips, all prove stalwart crops in cooler gardens.

Warmer temperatures allow you to push what you can grow in a temperate garden. We have many hardy, delicious edibles like apples, asparagus, and cherries, but with a sheltering wall, fence, or building in a warmer location you can experiment with kiwi fruit, feijoas, and Cape gooseberries.

Hot

Robust succulents and exotic edibles

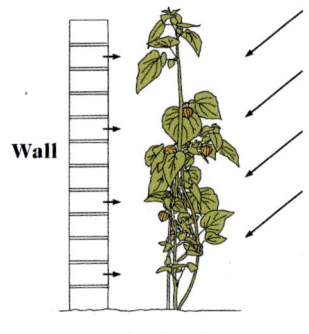

Edibles grown in built up areas such as towns and cities can benefit from a warm microclimate.

All plants, tender or hardy, have an optimum temperature range. Not being able to move around, plants have to adapt to temperatures, and, we can select these adaptations to our advantage. For most well-known edibles this optimum range is 10–30°C (50–86°F); in temperate climates, many of these will grow and yield perfectly adequately. But if global temperatures rise as predicted – by 2–4°C (35–39°F) by 2080 in the UK – gardeners may have to rely on heat-tolerant crops (or "thermophiles") like sweetcorn, sunflowers, soybeans, durum wheat, and sorghum. These plants can experience prolonged temperatures of over 30°C (86°F) with no deterioration in yield. This sounds futuristic, but as gardeners we are familiar with extreme thermophiles well-adapted to temperatures nearing 60°C (140°F), such as agaves and cacti. Corn tortillas, prickly pear, and agave syrup are all staples in Mexico.

Superior growth rate

Higher temperatures boost a plant's metabolism, resulting in faster growth. This supercharged growing means that gardeners in warmer locations have supreme space efficiency. Soils warm more readily in spring, so plots can be brought into cropping quickly. In wetter climates, the extra warmth boosts humidity, ideal for tropical edibles such as Malabar spinach, achocha, and chayotes.

Tropical edibles in urban areas

Location has a direct effect on temperature. Coastal plots have less extreme changes in temperature than inland, due to the moderating effect of the sea. We are increasingly choosing to live in urban areas rather than rural locations (in the UK this ratio is presently 83% : 17%). While associated growing spaces are smaller, city gardeners benefit from the "urban heat island" effect, where the sun's rays are absorbed by hard surfaces, and radiated out as warmer air. This raises temperatures significantly – by 10°C (50°F) in London – allowing tropical and subtropical edibles such as avocado to crop successfully.

LEARN TO READ YOUR PLOT

Heat-loving edibles from tropical climes will grow abundantly with sun and shelter, allowing the likes of Malabar spinach (far left), figs (left), chillies (below left), and aubergines (below) to crop with ease.

How to map cold and hot spots

Depending on its origins, each edible has a temperature range it performs best at. Too hot, and vital processes such as photosynthesis and transpiration stop; too cold, and essential plant tissues become irrevocably damaged. In temperate climates, we are lucky in that we are familiar with many edibles that tolerate freezing spells. With care and nurturing, we can cultivate crops from more tropical climes – and in doing so, we can enjoy edibles crops from growing environments all over the world.

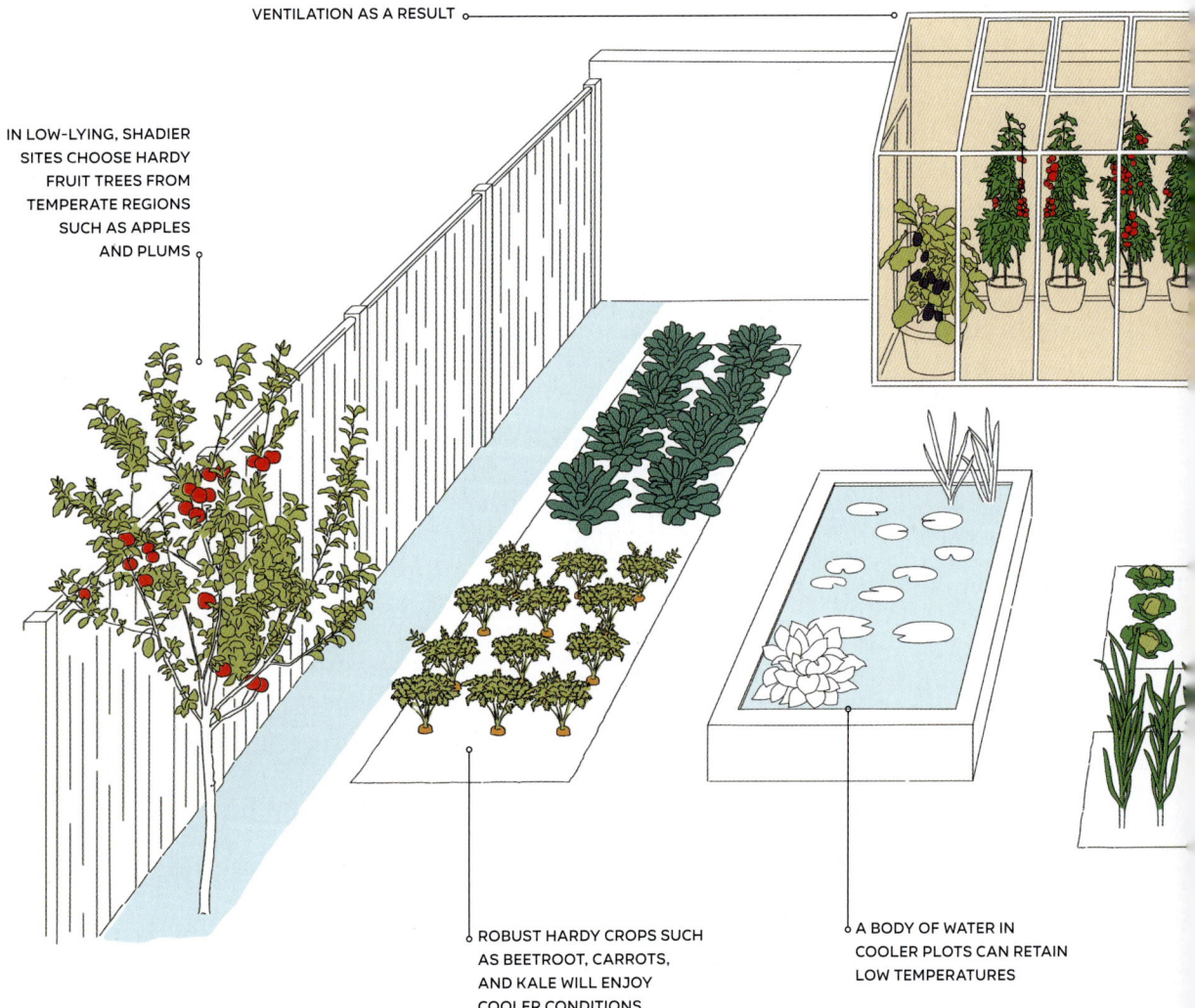

A GREENHOUSE IN FULL SUN HAS THE POTENTIAL TO BE INCREDIBLY WARM IN THE SUMMER AND MAY NEED VENTILATION AS A RESULT

IN LOW-LYING, SHADIER SITES CHOOSE HARDY FRUIT TREES FROM TEMPERATE REGIONS SUCH AS APPLES AND PLUMS

ROBUST HARDY CROPS SUCH AS BEETROOT, CARROTS, AND KALE WILL ENJOY COOLER CONDITIONS

A BODY OF WATER IN COOLER PLOTS CAN RETAIN LOW TEMPERATURES

As we experience more weather extremes, prolonged heatwaves and cold snaps are becoming the norm. By their sporadic nature they can catch us off guard, so it's best to be alert and prepare to safeguard harvests as much as possible.

Weather forecasts and apps give us at least 24 hours' notice of extreme temperatures, hot or cold. By understanding the general effect of high-pressure weather systems that bring settled weather, and low-pressure ones that encourage far more changeable conditions, we gain greater insight. Learning about the interaction of weather fronts and wind direction for your locality will enrich your knowledge further. For example, wet weather coming in from the west tends not to reach my home county of arid Essex, whereas bands of rain from the north or east generally do, and in winter, these are frequently combined with a fierce chill. Understanding your local climatic conditions in this way can positively influence your crop choices.

Frost pockets usually occur on sloping sites, either at the base of the gradient, or where cold air collects behind a solid structure, such as a wall or fence. Learn to identify these areas in your plot.

FRUITS THAT ENJOY LONG, HOT SUMMERS, SUCH AS FIGS, KIWIS, DESSERT GRAPES, AND PINEAPPLE GUAVAS, WILL THRIVE IN WARM LOCATIONS

A LARGE PAVED AREA IN FULL SUN HAS THE ABILITY TO HOLD A LOT OF HEAT, BOTH DAY AND NIGHT

A bracing walk on a windy day energizes and invigorates us, but how do plants feel being exposed to the breeze every day? Wind brings definite advantages, making coastal and hillside plots great places to grow your own.

Wind

Healthy harvests and strong growth

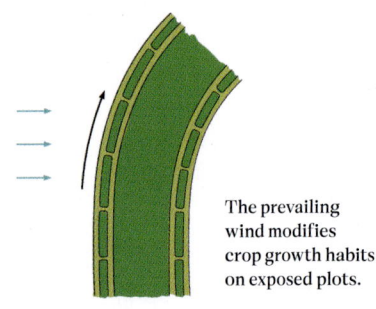

The prevailing wind modifies crop growth habits on exposed plots.

PLANTS ADAPT TO BREEZY CONDITIONS

Less pest and disease

Exposed gardens are perfect for robust crops like kale, leeks, and broad beans that hate being mollycoddled. Too much nurturing causes soft, leggy growth that's prone to pest and disease. Not so in windy locations.

A permanent breeze encourages robust, stocky, rot-proof growth, and air has no chance to become stagnant. Many fungi, grey mould, and late blight for example, thrive in humidity and stillness, and populations of aphids and whitefly prefer warm, sheltered environments. These all struggle to get a foothold when wind is ever-present.

Adaptations and benefits

Some crops thrive in a breeze. Sweetcorn is wind-pollinated and will provide full cobs with the wind dislodging pollen from male flowers. Other tall crops, such as Jerusalem artichokes, easily filter gusts, providing shelter for delicate edibles. Breezy areas are a useful environment for hardening off vegetable seedlings, and heathland crops such as lingonberries, alongside coastal salsify, and seakale, are perfectly happy buffeted about.

Plants also have wind-resistant adaptations. Leaf surfaces that are leathery (rosemary), waxy (globe artichoke), or narrow (onion and

LEARN TO READ YOUR PLOT

shallot) are naturally wind resilient. Many hardy plants become dormant in winter to hide from squally weather, by dying down or, as with deciduous plants, losing their leaves. Others such as lettuce, agretti, thyme, and summer purslane adopt a mounded shape so that gusts flow over them.

Minimize any limitations

Plots at high altitudes of more than 200m (656ft) above sea level may be several degrees cooler than low-lying neighbouring areas. This can mean a shorter growing season which, together with an exposed location, could reduce the range of crops suitable. In such cases, windbreaks and shelterbelts are invaluable, providing shelter for a distance of up to five times their height. They're preferable to solid walls and fences which can create turbulence on the leeward side.

Many crops have adapted to shrug off windy conditions. Seakale (far left) and globe artichokes (above left) both have tough, leathery foliage, whereas salsify (above) and lettuces (left) adopt a mound-like habit so that wind flows over them.

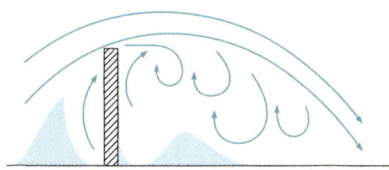

SOLID WALL

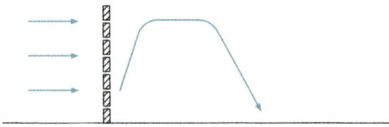

FENCE WITH SMALL GAPS

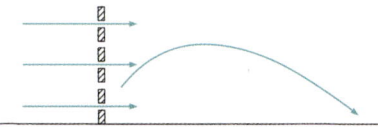

FENCE WITH LARGE GAPS

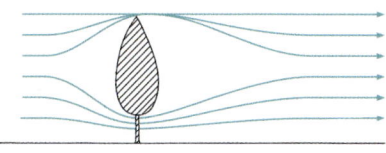

MATURE TREES

Solid walls can interact with wind aggressively, causing turbulence. Slatted fences and trees are preferable as they filter, rather than deflect, gusts.

Shelter takes the extremes out of gardening. Heat and wind are minimized, allowing you to cultivate lush, leafy edibles that would otherwise scorch. No longer will runner beans be torn to shreds, or potted peppers topple over, come the thunderstorms of high summer.

Shelter

Long seasons and lush leaves

(Right) A sunny, sheltered spot is ideal for ripening up potted chillies in autumn. (Below) Spring pollinators will be eager to visit peach blossom in shelter, and borderline hardy bay (below centre) can avoid winter scorch. (Far right) A sheltered spot supports the lax habit of tomatillo plants.

Borderline hardy crops, such as pomegranate, feijoa, and bay appreciate milder conditions. Get set for buttered new potatoes, tender spring cabbage, and baby carrots and turnips at the start of spring, as you encourage early crops under cloches and frames in your sheltered microclimate. You won't have to collect cracked or smashed cloches after a windy day (which can be heartbreaking and costly).

Valuable warmth at chilly times

A protected environment extends the seasons in spring and autumn. Slow-to-ripen tender crops like chillies, tomatoes, and melons, have ample time to yield. Autumn-sown broad beans are more likely to survive severe winters if sheltered from excess wet, and come spring, pollinating insects are eager to visit opening flowers, safe from chilly spring breezes. Add to this scene, freshly opened bright pink peach or nectarine blossom, and all this shelter sounds idyllic.

   If your sheltered plot is also mild and sunny, congratulations – wave goodbye to frost pockets damaging early-opening fruit flowers, and say

LEARN TO READ YOUR PLOT 39

hello to a long growing season that supports such delights as strawberries, kiwi fruits, tomatillos, and apricots. Use the area to cure harvested squashes, dry off lifted onions and garlic, or fully ripen potted chillies. If the sheltered areas in your garden are also shaded, cold air may settle there in winter, creating stubbornly icy pockets. Don't fret – earmark this area for supremely hardy edibles, or ones that bloom later in the season.

Urban gardening benefits

With their high-rise buildings, town and city gardens are more sheltered than rural plots. Embrace wall-trained fruit – ideal for compact areas where a plant's footprint is shadowed by its yield. Limb breakages from gales are minimal and evergreens like arbutus, gaultheria, and bay safe from leaf scorch and desiccation.

How to map wind and shelter

While windy sites can cause crops to hunker down, these environments also deter pest and disease, and encourage robust, rot-proof growth. However, a sheltered plot can be advantageous when growing exotic edibles that appreciate above average warmth. This nurtured location can also be incredibly handy for stretching the seasons, leading to longer harvest periods in spring and autumn.

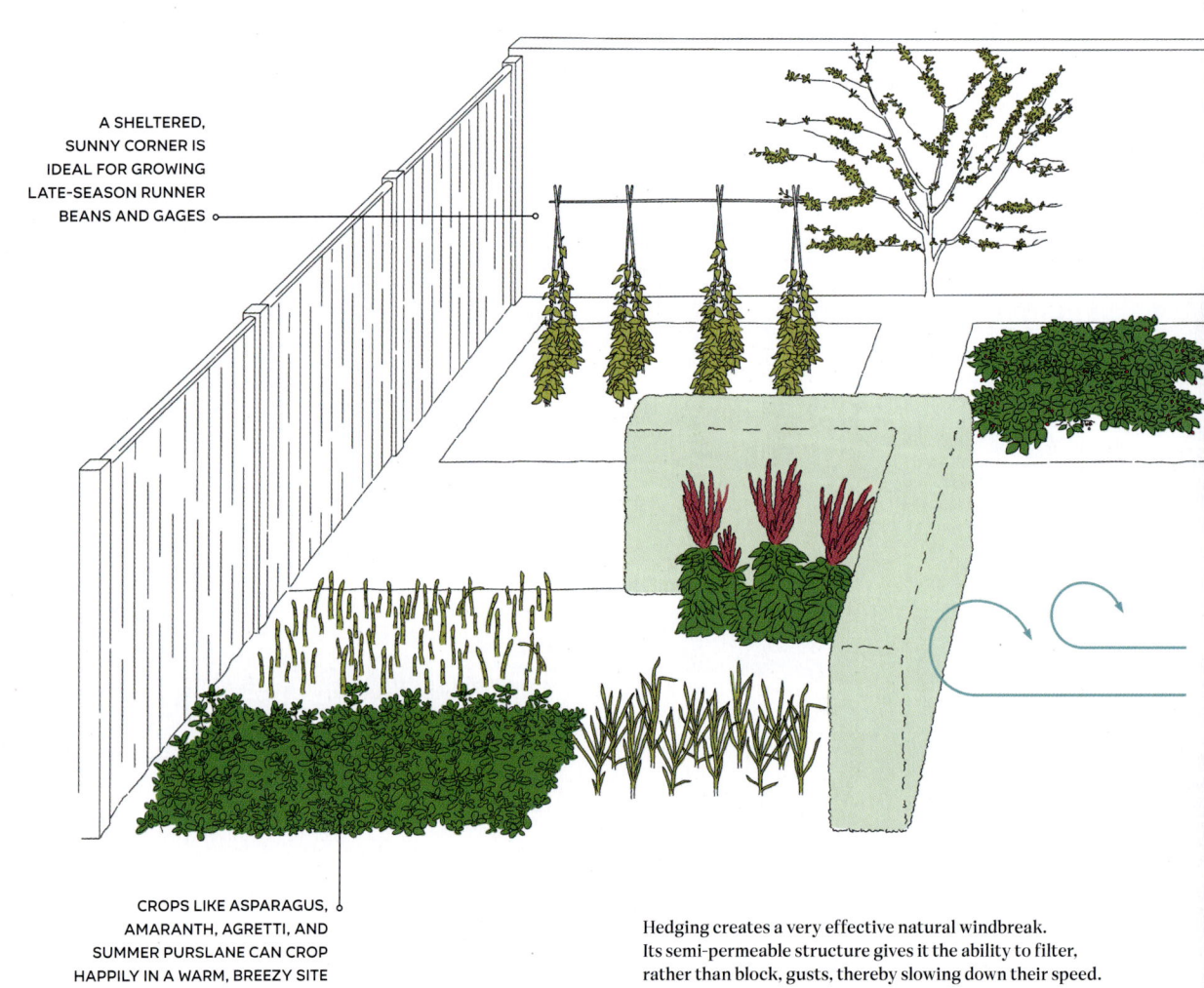

A SHELTERED, SUNNY CORNER IS IDEAL FOR GROWING LATE-SEASON RUNNER BEANS AND GAGES

CROPS LIKE ASPARAGUS, AMARANTH, AGRETTI, AND SUMMER PURSLANE CAN CROP HAPPILY IN A WARM, BREEZY SITE

Hedging creates a very effective natural windbreak. Its semi-permeable structure gives it the ability to filter, rather than block, gusts, thereby slowing down their speed.

LEARN TO READ YOUR PLOT

While wind is generally governed by the wider local geography, shelter can be created within any individual plot. Aim to identify the windward (exposed) and leeward (sheltered) aspects of your growing area. You may well find seasonal variations due to autumn gales, or even daily ones near coastal areas.

Man-made structures can create turbulence too. Narrow passageways often create channels where wind is forced through at speed. Solid walls force air upwards, and then a strong downdraught is generated on the leeward side.

Shelter is often found in plot corners, where the wind finds it impossible to access at speed. Create more by erecting well-placed fences or walls, or by planting living windbreaks. Conversely, stagnant sites can be opened up by removing man-made divisions, or by pruning back natural ones.

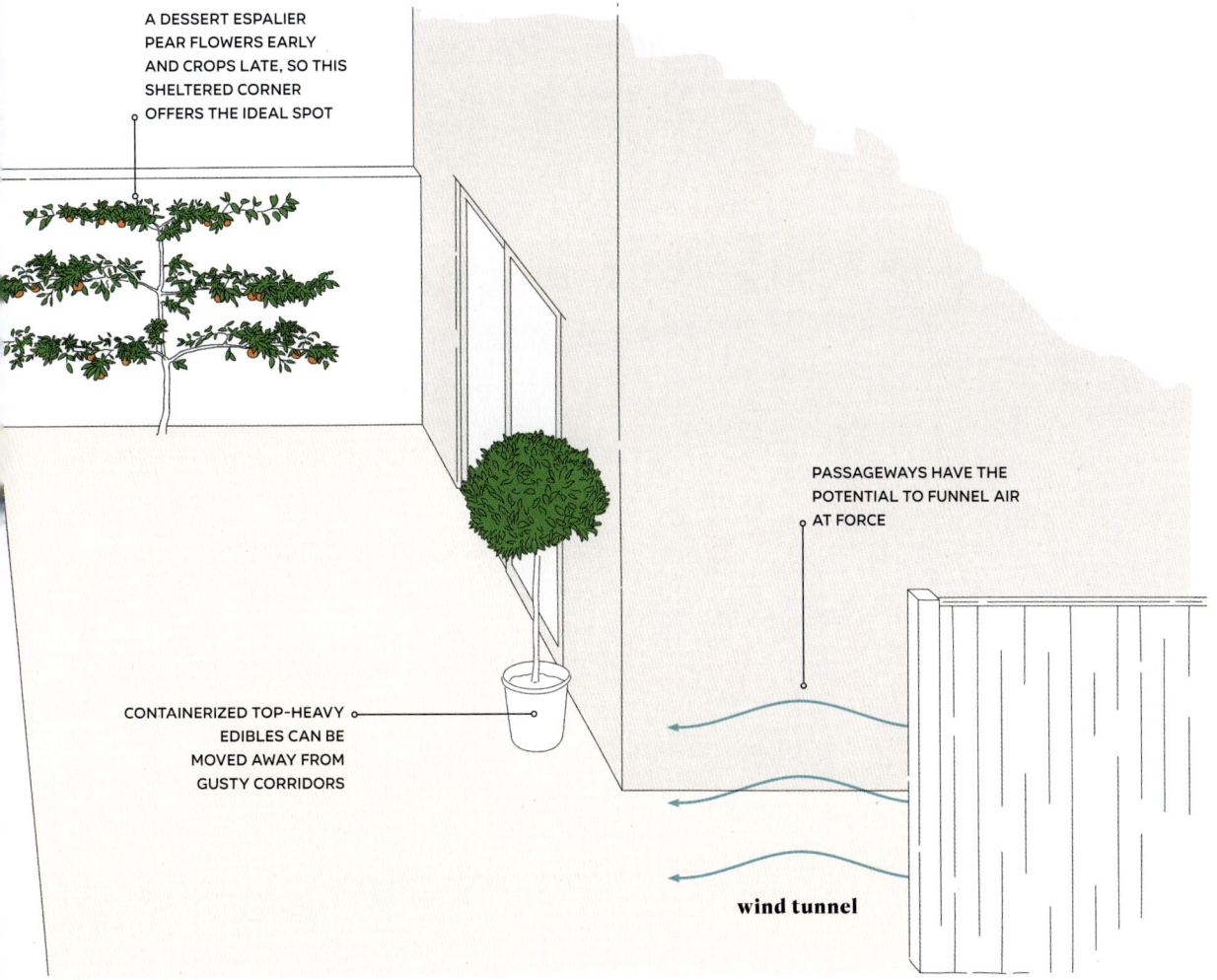

A DESSERT ESPALIER PEAR FLOWERS EARLY AND CROPS LATE, SO THIS SHELTERED CORNER OFFERS THE IDEAL SPOT

PASSAGEWAYS HAVE THE POTENTIAL TO FUNNEL AIR AT FORCE

CONTAINERIZED TOP-HEAVY EDIBLES CAN BE MOVED AWAY FROM GUSTY CORRIDORS

wind tunnel

Keen grow-your-owners often earmark a specific area in their garden for an edible plot. These kitchen gardens may be in miniature but they can prove to be hugely productive spaces. By design, they offer many different growing environments for a wide range of edibles.

Case study:
A devoted edible garden

1 Wind tunnel in full sun caused by narrow passageway further up the garden: Robust plants are positioned here such as apples, leeks, blueberries, and hardy salads.

2 Shady, damp corner against a north-facing fence: Fruits happy in low light levels, such as currants, alpine strawberries, and rhubarb, are planted here, along with leafy veg like sorrel and land cress.

3 Sunny, moist beds: These areas are warm and open, and offer excellent conditions for tender, hungry crops like winter squashes, French beans, Florence fennel, shiso, and sweetcorn.

4 Narrow beds in full sun: Home to the most drought-proof edibles, such as globe artichokes, agretti, amaranth, and seakale. The fence behind is planted up with sweet cherries, figs, and peaches.

5 Sheltered, sunny corner: This area is ideal for forcing early crops in spring. Baby spinach and strawberries are planted under cloches. In autumn, pots of chillies and aubergines are placed here to ripen up.

This square plot (opposite) contains sunny beds, shadier areas, sheltered corners, and an exposed wind tunnel.

(Below) This sunny corner is perfect for growing strawberries and chillies. (Bottom) The shady wall supports redcurrants and garlic mustard.

CASE STUDIES 43

Large gardens flanked by fences and hedges are frequently associated with detached or semi-detached family homes. Often juggling multiple uses, these plots offer a wide range of environments for crops, too. You can easily enhance garden structures and flowerbeds with edible plantings.

Case study:
A larger detached property

1 Sheltered seating corner: Ideal environment for tropical-themed potted edibles such as cannas, dahlias, amaranth, and sweet potatoes. Shadier sides can be home to tubs of hostas and edible ferns.

2 Patio area: Sunny areas can support a potted collection of culinary herbs for the kitchen, such as coriander, oregano, sage, bay, and basil. Shade can support leafy salads in growbags, along with potted cane fruits.

3 Woodland margins: The shady, cooler beds here will make an excellent location for spinach, Swiss chard, hostas, and redcurrants. You could even establish a mushroom "loggery".

4 Shrub border: Can be interplanted with woody edibles happy in part shade, such as barberry, honeyberry, and dwarf raspberries. Plant alpine strawberries for ground cover.

5 Greenhouse and raised beds: This devoted, sunny area can be highly ornamental as well as productive. Use it to support traditional crops such as French beans and courgettes, alongside greenhouse tomatoes, peppers, melons, and cucumbers.

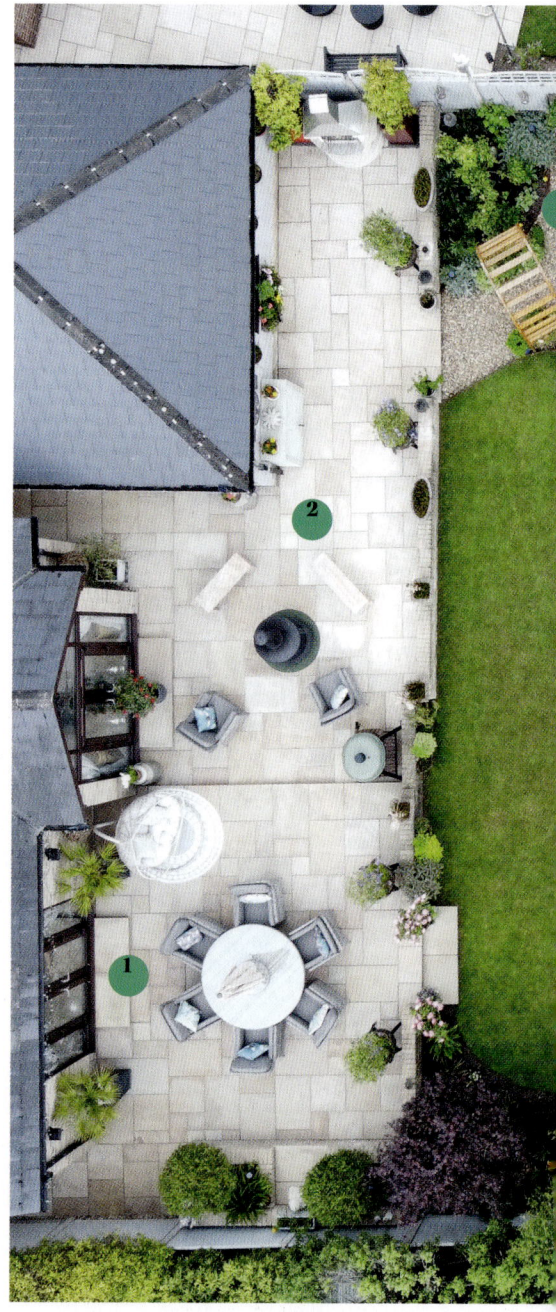

This large garden is flanked by fence panels and contains a paved area for socializing. The presence of both sun and shade allows for a diverse choice of crop.

CASE STUDIES **45**

(Top) This patio suntrap could easily support a selection of potted Mediterranean herbs.

(Above) The greenhouse area offers extra shelter for heat lovers like tomatoes, basil, and aubergines.

Village, town, and city dwellers often have compact plots with multiple features. These offer the perfect microclimate for many edibles, and weeds are easier to control when growing in containers or smaller beds. Frequently sheltered, these urban edible gardens can house more tender crops.

Case study:
A compact garden

1 Sunny, sheltered patio and walls: These are ideal for supporting large troughs planted with climbers such as dessert grapes, kiwi fruit, and figs. Cordon tomatoes in pots also readily ripen in the sunshine.

2 Sunny paving by walls: Exotic potted crops such as ginger, dahlia, lemongrass, pomegranate, chillies, aubergines, and edamame will thrive here. An avocado tree can be grown in a large tub.

3 Fence in shade: This vertical growing area could be highly productive if clad in redcurrants, Japanese wineberries, and blackberries.

4 North-facing paving: This sheltered, shady area is ideal for large tubs of rhubarb, along with an attractive display of potted hostas and edible ferns, plus windowboxes of Oriental leaves.

(Right) This shrub bed in part shade would be the ideal spot for berrying perennials, such as Chilean guavas, dwarf raspberries, fuchsias, and serviceberries.

(Far right) This garden building is in sun and shelter – it could make the perfect support for a grape or kiwi fruit to scramble over.

CASE STUDIES 47

1 Open beds in full sun: Much of this plot is exposed to the elements. Being adjacent to a field, wind blows freely. Wind-tolerant, hardy veg such as carrots, onions, and leeks will grow well here.

2 Compost area: Ample organic matter, a water supply, and full sun make this a fertile area. Ideal for space- and nutrient-hungry rambling pumpkins and butternut squashes, and ridge cucumbers.

3 Large water tanks: The sides in full sun are perfect for a sprawling grapevine. Tank walls facing the large tree and hedge are in shade, making them a great spot for training loganberries, gooseberries, and currants.

4 Large tree: This has extensive roots and casts summer shade. Blackcurrants, worcesterberries, and elder provide a miniature "forest garden", and a raised bed grows garlic mustard, sweet cicely, and horseradish.

This allotment offers a sizeable and exposed growing area. The mature tree casts shade and will compete with the edibles for moisture.

Case study:
An allotment plot

For many gardeners, renting an allotment is the easiest and most cost-effective way of having access to a large growing space. Generally rectangular in shape, they often come with inherited structures, such as sheds, fruit cages, and even mature trees, which can determine your crop choices.

CASE STUDIES 49

The importance of soil

Understanding soil in order to grow food, anywhere

When it comes to soil, the vast majority of gardeners possess a predominant soil type. You might garden on sticky clay, flinty chalk, or light sand, and attempts to adjust it significantly are impractical. Adding excessive amounts of sulphur, lime, or fertilizer onto your plot is not only costly to you, but irresponsible to the environment.

My coastal plot is on very sandy soil, yet residents just five minutes inland by car garden on heavy clay. Over the years, I've learnt what performs brilliantly in my garden with little intervention from me. Peaches, figs, asparagus, globe artichokes, tomatoes, and basil all revel on my plot. If I want to grow a rose, I'd be wise to choose *Rosa rugosa*, which originates from coastal dunes. If I want effortless home-grown summer salads, I'll sow agretti, summer purslane, and New Zealand spinach. But I also love to grow raspberries, which succeed in cool, moist conditions. Planted in part shade, and mulched with home-made compost each spring, I keep the soil and roots cool and moist – inexpensively and minimally.

Unlike sun, shade, wet, and dry – all of which can be found in any given space – the soil in our garden is homogeneous and we shouldn't need or want to significantly

THE IMPORTANCE OF SOIL

Your specific soil type has been created over millennia, and we should choose crops that have evolved to thrive in it.

change it. By understanding its properties, we can garden holistically, choose crops that perform fantastically in our conditions, and intervene minimally. That way, we can grow an exciting and diverse range of edibles.

What is soil?

Created over millennia, soil is a mixture of weathered rock and organic matter, the composition of which varies depending on where and how on the planet it was formed.

Within the British Isles alone, Scottish gardeners at altitude will be familiar with "young", seemingly unweathered soil, comprised of rock and thin, peaty bogland moistened by underground springs. Gardeners on the breezy north Norfolk coast may not realize their luck, forks sliding into loose, sandy soils deposited via a system of estuaries known as "The Wash". Residents of the Yorkshire Wolds and Surrey Downs would envy such soil, as they instead come across large chunks of flint and chalk, created 300 million years ago via plankton deposits in tropical seas. So would gardeners on heavy Essex and London clay, which sticks to fork prongs like glue, yet is amazingly fertile due to the settling of nutrient-rich algae, over 50 million years ago.

Unfathomable this may sound, but remember, edibles evolved for millennia, globally, to thrive in specific soils like these.

What kind of soil do you have?

This question may sound daunting to a new gardener, but it's easy to discover the answer with a little research. Ask locally – a seasoned gardener should be happy to offer advice. Look for "indicator" plants growing nearby; rhododendrons enjoy acidic soil, lavenders thrive in chalky areas, nettles like rich soils, and clovers enjoy those low in nitrogen. When you feel the soil, is it gritty (sandy soil) or sticky (clay soil)? Digging a soil "profile pit" – a hole at least 50cm (20in) across and 1m (3ft) deep – exposes topsoil and subsoil, potentially revealing chalk, clay, or sand. If it puddles with water this may indicate waterlogging.

Soil exceptions

You may have bought in topsoil to fill raised beds or re-landscape your plot. If it is good quality it will be screened for stones and weeds, neither too chalky, sandy, peaty, or high in clay, and may come with a nutrient and pH analysis. Your outdoor space may be a balcony, patio, or courtyard where all growing is carried out in containers – if so, it's important to choose the right compost for your crops.

Soil-plant interactions

Soil characteristics can interact with your growing "zones". For example, crops on exposed sites can struggle to anchor themselves on shallow peat or chalk, but can root extensively on deep, loamy plots. Thankfully, soil types can coincide with zone types: peaty soils are often wet, whereas sandy and chalky soils are predominantly dry. To master these interactions so that we can grow effortless edibles, it helps to understand a few factors.

Soil texture and structure
Is your soil predominantly clay, sand, loam, chalk, or peat? The following table gives examples of how soil texture influences crop choice:

SOIL TEXTURE
CLAY
SAND
LOAM
CHALK
PEAT

CHARACTERISTICS	BEST FOR	GARDENING IMPLICATIONS
Slow to warm in spring, drains poorly, waterlogged in soggy winters/baked hard in dry summers, nutrient rich	Excellent for hungry cabbages and squashes	Not ideal for forcing early crops in shelter, or growing fleshy-rooted/borderline hardy edible perennials on moist soils, harder to dig
Quick to warm in spring, sharply drained, not moisture retentive, nutrient low	Ideal for stocky, maritime crops like agretti and seakale	Perfect for sheltered cloched areas of early edibles, more successful overwintering of tender plants in cold locations, easy to dig
Quite fertile, can be a bit "tacky" after rain	Ideal for edibles preferring direct-sowing like turnips, carrots, parsnips, or slow, steady moisture and nutrients like bulb fennel, and beetroot	Nothing too extreme, therefore desirable to own, easy to dig
Topsoil can be shallow, very alkaline (see pH, page 57), drains well	An excellent base for many niche crops such as crab apples, mulberries, perpetual spinach, and perennial kale	Useful for overwintering less hardy perennials, can be "chunky" to dig, especially if it contains flint
Retains moisture well, can be waterlogged but also drains easily, potentially shallow and rocky, very acidic (see pH, page 57)	Ideal for acid-loving "ericaceous" fruits like blueberries, cranberries, lingonberries, and gaultheria	Easy to dig if drained

Choosing appropriate crops for your soil type encourages healthy root growth, leading to better harvests.

Looking after soil structure

Maintaining good soil structure is important as situations such as compaction and waterlogging can cause damage and limit crop choices. Don't walk on or work soil when it's frozen or waterlogged as compaction can result, where roots can find it very difficult to penetrate. Ease permanent waterlogging with a soakaway (essentially a large gravel pit) or drains and ditches on larger areas. Temporary waterlogging can be alleviated by adding bulky organic matter like garden compost to the topsoil.

Soil fertility and nutrition

In nature, major plant nutrients such as nitrogen, phosphorus, and potassium are continually cycled between a diverse range of plants and the soil beneath them. When gardening, we generally plant beds of single crops and harvest them. Harvesting breaks that cycle, depleting soil nutrients. This practice alone means that for good yields it's beneficial to supplement our gardens with nutrients.

Soil texture also plays a significant role in nutrient availability. For example, sandy soils don't retain nutrients as well as clay soils. Dry soils have little free soil water – where most nutrients are dissolved for plant uptake – and wet soils can have nutrients washed out of them. A solution for gardeners with both these soil types is to add bulky organic matter like garden compost or rotted animal manure.

A living soil

As soon as soil starts to form, it becomes a precious habitat teaming with life – and a healthy, living soil promotes healthy, robust plants. We are all familiar with larger soil residents such as voles, earthworms, millipedes, and ants, but equally if not more important are microscopic springtails, nematodes, mites, bacteria, fungi, and algae. Between them they break down soil, releasing valuable nutrients in the process and making these available to plants.

These creatures are happier in well-aerated soils, struggling to survive in extreme wet or dry. Very high or low soil pH affects the life forms that can survive. Luckily, you can boost soil health with garden compost.

Soil pH

This explains how acid or alkaline your soil is. Like conditions such as dryness and exposure, soil pH can also govern what will grow well on your plot.

Measured on a scale of 1–14, a neutral pH is 7, anything below that is acid, anything above alkaline. Sandy soils lose calcium as they age, gradually becoming more acidic. Peat-based soils too, are often acidic. Chalky soils conversely, are continually replenished with calcium so remain alkaline. Soil pH has a significant effect on nutrient availability, and at extremes elements like nitrogen, magnesium, and iron can be "locked" onto soil, making them unavailable to certain plants. Some edibles evolved to thrive at pH extremes, like blueberries on acid soil. Choosing these reduces our need to adjust pH. Some crop diseases are also strengthened by pH – potato scab is more prevalent on alkaline soils, and brassica clubroot can be widespread on acidic plots.

Be mindful of soil properties, understand their interactions with edibles, keep intervention to a minimum, and choose appropriate crops, and you can support optimum crop growth on any plot.

BLUEBERRY
Blueberries have evolved on, and therefore enjoy, acidic soils.

BEETROOT
Many crops, including beetroot, are happiest in a soil with a pH of 6–8.

CABBAGE
Alkaline plots can deter destructive soil diseases, such as clubroot on cabbages.

Matching crops to zones

In temperate gardens, locating the sunniest corner of your plot, protected from the ravages of wind, opens up the opportunity to cultivate all manner of tender, exotic edibles – it's the closest some of us get to owning a greenhouse. Heat is abundant, so sugar levels of fruits like apricots, peaches, cherries, and strawberries will all naturally and effortlessly rise. Tender leaves, otherwise battered by gales, can unfurl to their full potential, leading to trugs full of velvety Malabar spinach, basil, and beans. And let's not forget plump sweet potatoes, tomatoes, and peppers, packed full of umami flavour. In essence, you've discovered the sweet spot for any grow-your-own fanatic.

Zone 1
Sunny and sheltered

ZONE 1: SUNNY AND SHELTERED

Peaches, nectarines, and apricots

Prunus species

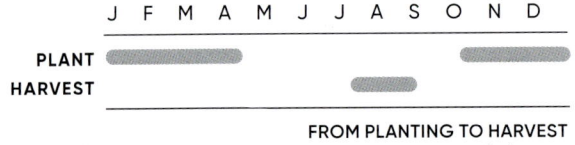

FROM PLANTING TO HARVEST
1–4 years

VARIETIES
Peach 'Rochester' – vigorous, productive tree, late flowering

Peach 'Saturne' – a peento peach, flat fruits, delicious white flesh

Nectarine 'Lord Napier' – reliable, vigorous, productive variety

Apricot 'Tomcot' – modern variety, vigorous, abundant, late flowering

ORIGINS
Temperate Asia

ALSO GROWS IN
Zones 2 and 3 – long, hot summers are key

HARDINESS
Hardy, flowers vulnerable to frost

LIFE CYCLE
Perennial

YIELD
★★★★½☆☆☆☆☆

EASY TO GROW
7/10

Many are surprised that temperate climates can produce delicious peaches, nectarines, and apricots but with sufficient sun it's absolutely possible. Ideally fan-train trees in a sunny spot – in full flower this is a beautiful sight.

Growing essentials

Planting A sunny position and well-drained soil dug over with garden compost or well-rotted manure. To fan-train see pages 74–75.

Pruning Prune trees in spring and summer as winter pruning invites disease. Initial years consist of tying in framework stems. At maturity remove a proportion of old wood and tie in new stems.

Harvesting Harvest fruits when soft and heavy with juice. If wasps show interest, pick fruits slightly earlier to ripen on a sunny windowsill.

Secrets of success

Being self-fertile, only one tree is needed for fruiting.

A winter chill is essential for good flowering. Choose late-flowering varieties for cool climates.

A sheltered position invites bees to pollinate early peach blossom, leading to sizeable harvests.

Prevent peach leaf curl by cloaking trees in clear, reusable polythene from January until April (see page 211).

Cover open blooms if frost is forecast. Use a double layer of fleece, holding it off blooms with canes.

Hand-pollinate flowers with a soft paintbrush if bees are hindered by fleece.

A succession of new stems is key for plants to crop well. Prune out any old wood annually.

ZONE 1: SUNNY AND SHELTERED

Sweet cherries
Prunus avium

	J F M A M J J A S O N D
PLANT	▬▬▬▬▬ ▬▬▬
HARVEST	▬

FROM PLANTING TO HARVEST
1–4 years

VARIETIES
'Sunburst' – self-fertile, large almost black fruits, mid-season

'Celeste' – self-fertile, compact, early season

'Stella' – reliably self-fertile, heavy-cropping, mid- to late season

'Lapins' – self-fertile, heavy mid-season yields, tasty fruits

ORIGINS
Widely distributed tree, enjoys warmth but not waterlogging

ALSO GROWS IN
Zone 3

HARDINESS
Fully hardy

LIFE CYCLE
Perennial

YIELD
★★★/☆☆☆

EASY TO GROW
7/10

Growing top quality, delicious cherries is an easy art to master if trees are given the right position and bird protection. Adequate sunshine secures the sweetest of fruits, and trees can be kept to a manageable size with pruning. Enjoy show-stopping blossom displays and autumn leaf colour, too.

Growing essentials

Planting Plant in a sunny, sheltered position in well-drained soil, dug over with garden compost or well-rotted manure. Free-standing or trained against a wall or fence.

Pruning Initial years consist of building up stem framework. Prune established trees in summer, after fruiting. At maturity remove a proportion of old wood and tie in stems of trained trees.

Harvesting Cherries taste best fully ripened. Harvest once deeply coloured for best flavour and juice, picking by the stalks to avoid bruising.

Secrets of success

Cooking cherries, like *Prunus cerasus* 'Morello', crop well in shade.

Prune young trees mid- to late spring, before fruiting, to build up shape.

Avoid pruning October to March to deter bacterial canker and silver leaf.

Maggoty fruits may indicate spotted wing drosophila – cloak fruits in fine mesh.

Sweet cherry trees revel in sunshine, and their fruits are deliciously juicy once fully ripe.

Fan training offers an excellent way to grow trees against a wall or fence (see pages 74–75).

Skins can split if they get wet when ripening. Cloak fruits in reusable polythene.

ZONE 1: SUNNY AND SHELTERED

Basil
Ocimum species

	J	F	M	A	M	J	J	A	S	O	N	D
SOW					▬	▬						
TRANSPLANT						▬						
HARVEST							▬	▬	▬			

FROM SOWING TO HARVEST
10–12 weeks

VARIETIES
'Aroma 2' F1 – robust and vigorous, soft, delicious leaves

'Siam Queen' – spicy Thai basil, purple flowers

'Dark Opal' – striking purple stems and leaves

African basil – or tree basil, a perennial species, *O. gratissimum*

ORIGINS
Mainly tropical Asia, in sunny, hot areas with little summer rainfall

ALSO GROWS IN
Zones 2 and 8 – excellent windowsill plant

HARDINESS
Not hardy – killed by frosts

LIFE CYCLE
Predominantly annual, though perennial forms exist

YIELD
★★★★½/☆☆☆☆

EASY TO GROW
7/10

Basils 'Red Rubin' (left), 'Aroma 2' (centre), and 'Queen of Sheba' (right) will all thrive in warmth and shelter.

A staple Mediterranean ingredient, with a distinctive aroma, basil loves sunny, free-draining beds. It is well suited to pot cultivation, with an abundance of varieties, each with subtle flavour variations.

Growing essentials

Sowing Sow 3–4 seeds clustered together, in modular cells, under cover. Once well emerged with roots filling cells, transplant clumps into 9cm (3½in) pots.

Planting Grow on somewhere frost-free then plant out once the risk of frost has passed, spaced 30cm (12in) apart, or in 30cm (12in) pots.

Harvesting Keep houseplants in pots. Harvest as and when needed, sowing batches all summer. Harvest by pinching out tips to keep bushy.

Secrets of success

Keep seedlings above 18°C (64°F), don't overwater to avoid damping off disease.

Grow as a microleaf for harvestable leaves in just 3 weeks.

Propagate with cuttings – place stems in a jar of water until roots emerge.

Basil flowers are edible but their production limits foliar growth, so remove them for maximum leaves.

Pick gluts of leaves from their stems, wash, bag up, and freeze.

Grow indoors year-round. Sow in small pots in batches, or plant cuttings.

ZONE 1: SUNNY AND SHELTERED

Runner beans

	J	F	M	A	M	J	J	A	S	O	N	D
SOW					▇							
TRANSPLANT							▇					
HARVEST								▇▇▇				

FROM SOWING TO HARVEST
16–20 weeks

VARIETIES

'Firestorm' – excellent, red flowers, pods well in hot and cold

'Snowstorm' – white-flowered, sets pods very well, even in extremes

'Moonlight' – high quality, white flowers, self-fertile, reliable

'Hestia' – attractive dwarf form, red and white flowers, ideal for pots

ORIGINS

Central America. Breeding has improved drought and cold resistance considerably

ALSO GROWS IN

Zone 3, and zone 5 if plants can reach sun

HARDINESS

Not hardy – killed by frosts

LIFE CYCLE

Perennial, grown as an annual

YIELD

★★★★⯪☆☆☆☆

EASY TO GROW

8/10

Runner beans are a stalwart crop for good reason – pods keep on appearing with regular harvesting, from July into late autumn. Pod set and tenderness have improved in quality with breeding, with compact types ideal for containers.

Growing essentials

Sowing Sow seeds 3cm (1¼in) deep, in individual pots under cover in spring.

Planting Once seedlings emerge, harden off, then plant alongside beanpoles set at 40cm (16in) apart, or in rows 40cm (16in) apart. Protect young plants from winds, and water well until established. Sow seeds outdoors once the risk of frost has passed. Climbing forms twine up supports unaided.

Harvesting Attractive flowers appear in summer, followed by pods growing quickly in warm, moist conditions. Harvest regularly, whilst young and tender.

Secrets of success

Avoid sowing early as seedlings struggle in chills.

A warm propagator set to 18–24°C (64–75°F) gives the best germination.

Dig garden compost and bonemeal into beds to sustain these vigorous plants.

Climbing forms grow high – ensure supports are strong and tall (3m/10ft).

Overhead foliar feed of seaweed is useful for boosting plants mid-season.

Pick all pods so that more form – don't allow pods to develop swelling beans.

British-bred 'Firestorm' will reward you with a phenomenal succession of tender pods when given the right conditions.

ZONE 1: SUNNY AND SHELTERED

Strawberries
Fragaria × ananassa

	J	F	M	A	M	J	J	A	S	O	N	D
PLANT			▬	▬	▬	▬	▬	▬	▬	▬	▬	
HARVEST					▬	▬	▬	▬	▬			

FROM PLANTING TO HARVEST
8–32 weeks

VARIETIES
'Vibrant' – heavy yields, juicy, flavoursome, early season

'Cambridge Favourite' – tried and tested, mid-season, tasty and reliable

'Pegasus' – late season, well-flavoured, disease resistant

'Albion' – perpetual, large fruit clusters, good disease resistance

ORIGINS
Derived from two American species cross-bred over time

ALSO GROWS IN
Zone 3

HARDINESS
Hardy

LIFE CYCLE
Perennial

YIELD
★★★/☆☆☆☆

EASY TO GROW
8/10

Whether plucked from the plant and warm from the sun, or combined with cream, sugar, and meringue, home-grown strawberries are one of the real treats of summer. Compact, fuss-free, and quick to crop, they're deservedly popular with gardeners.

Growing essentials

Planting Plant runners into weed-free beds enriched with well-rotted organic matter, in soil neither frozen nor waterlogged. For pots, use multi-purpose compost, placing plants 20cm (8in) apart.

Maintaining Keep young plants well watered, but not soggy, until established. Add a balanced liquid feed, switching to high-potash when flowering or fruiting.

Harvesting Net or cage fruits from birds. A layer of straw keeps soil-grown fruits clean. Harvest once richly coloured for the best flavour. Pick via the stalk to avoid bruising.

Strawberries appreciate being bathed in sunshine – their flavour is best when eaten still warm from the sun.

Secrets of success

Seeds are available too – sow under cover, in spring warmth, to transplant out later.

Force early varieties in a greenhouse or conservatory. Gently run hands over open flowers to pollinate them.

Propagate by pegging healthy runners down in soil or small pots until they root.

Young plants crop best their second and third years – keep circulating new stock via runners.

Some varieties like 'Toscana' and 'Summer Breeze Rose' bear attractive pink flowers instead of white.

Plants are available in small pots or plugs in spring and bare-root runners in summer.

ZONE 1: SUNNY AND SHELTERED

Tomatoes

Solanum lycopersicum

	J	F	M	A	M	J	J	A	S	O	N	D
SOW				▬	▬							
TRANSPLANT						▬	▬					
HARVEST							▬	▬	▬	▬		

FROM SOWING TO HARVEST
18–24 weeks

VARIETIES

'Rubylicious' F1 – attractive cordon, red cherry toms, blight resistance

'Losetto' F1 – quick cropping, trailing dwarf variety, red cherry toms, blight resistance

'Rose Crush' F1 – vigorous cordon, pink beefsteak fruits, blight resistance

'Nagina' F1 – cordon with plum-shaped fruits, blight resistance

ORIGINS
Tropical highlands Mexico and the Americas

ALSO GROWS IN
Zones 2 and 3 – if sufficiently warm

Zone 8 – compact varieties as houseplants

HARDINESS
Not hardy – killed by frosts

LIFE CYCLE
Perennial, grown as annual

YIELD
★★★★/☆☆☆☆

EASY TO GROW
7/10

Embraced by a vast range of cultures due to their flavour, diversity, and easy cultivation. Full sun guarantees bumper yields, and blight resistance breeding is transforming outdoor cultivation.

Growing essentials

Sowing Sow seeds 5mm (⅛in) deep, 10mm (½in) apart, in pots under cover. Once seedlings are large enough to handle, transplant into individual 9cm (3½in) diameter pots, grow on under cover.

Planting Transplant outside as the first flowers appear, after risk of frost has passed. Grow cordons in 30cm (12in) pots, growing bags, or borders. Dwarf types are ideal for baskets and smaller containers.

Harvesting Train cordons up twine or canes, removing sideshoots. Water well. Harvest fruits when ripe.

Secrets of success

Set a propagator at 18–22°C (64–71°F) for optimum germination.

Sow crops destined for greenhouse harvests in early February – cordons are space-efficient and productive.

Buy plants if you don't own a propagator – grafted forms are heavy cropping.

Give high-potash liquid feed when flowers appear, every 6–10 days.

Initiate ripening of the last tomatoes by reducing watering, pinching out cordon main stems, and removing immature fruits.

Ripen fruits by allowing them to colour up fully on a sunny windowsill – this boosts sugar levels.

Tomatoes 'Rubylicious' (left), 'Black Moon' (centre), and 'Losetto' (right) all enjoy a sunny, sheltered position outside.

ZONE 1: SUNNY AND SHELTERED

Malabar spinach

Basella species

	J	F	M	A	M	J	J	A	S	O	N	D
SOW					●							
TRANSPLANT						●						
HARVEST							●	●	●	●		

FROM SOWING TO HARVEST
15–18 weeks

VARIETIES

Basella alba – light green leaves, white flowers

B. rubra – deep green leaves, red flowers

There are presently no named varieties

ORIGINS

Known as Ceylon or Indian spinach, found in moist, tropical Asian forests

ALSO GROWS IN

Zones 3, 5, and 6 – if sufficiently warm

HARDINESS

Not hardy – killed by frosts

LIFE CYCLE

Perennial, but grown as an annual

YIELD

★★★★½/☆☆☆☆

EASY TO GROW

8/10

This vigorous, twining vine yields phenomenally in ideal conditions. Its large, kidney-shaped leaves taste tender and velvety raw, and can be wilted down as a nutritious green vegetable. A forest dweller, it grows in shade, too.

Growing essentials

Sowing Sow seeds 2cm (¾in) deep, singly in large modules under cover. Once seedlings reach 6–8cm (2½–3in), harden off, ready to plant out once risk of frost has passed.

Planting Apply garden compost generously to the planting area. Space plants 30cm (12in) apart, training up canes, beanpoles, or strong twine, at least 2m (6½ft) tall. Initially twining upwards – established plants will produce sideshoots.

Harvesting Begin harvesting leaves once well established. Continue harvesting until plants are killed back by frosts.

Secrets of success

For excellent germination set propagator to 15–20°C (59–68°F).

Provide as much warmth, shelter, and moisture as possible for maximum yields.

If you have a large glasshouse, plants can grow up to 4m (13ft) tall.

Leaves stay tender when mature, showing no signs of becoming fibrous.

Both *Basella alba* (top) and *B. rubra* (above) can grow at an astonishing rate, given warmth.

No vulnerability to pests or diseases – a great crop for organic gardeners.

Save seeds from mature plants – collect in autumn and dry off.

ZONE 1: SUNNY AND SHELTERED

Peppers
Capsicum species

	J	F	M	A	M	J	J	A	S	O	N	D
SOW			▬	▬	▬							
TRANSPLANT						▬						
HARVEST							▬	▬	▬	▬		

FROM SOWING TO HARVEST
24–32 weeks

VARIETIES
Sweet pepper 'Thor' F1 – tapered fruits up to 25cm (10in)

Sweet pepper 'Popti' F1 – stocky, with many small, red fruits

Chilli pepper 'Jalapeno' – compact, bullet-shaped, mild fruits

Chilli pepper 'Quickfire' – compact, abundant hot red fruits

ORIGINS
Central and southern America, thriving in hot, sunny areas, frequently with minimal rainfall

ALSO GROWS IN
Zone 8 – compact varieties make excellent houseplants

HARDINESS
Not hardy – killed by frosts

LIFE CYCLE
Perennial, predominantly grown as an annual

YIELD
★★★★/☆☆☆☆

EASY TO GROW
8/10

Peppers 'Popti' (left), 'Hot Lemon' (centre), and 'Habanero Orange' (right) adore a sunny location.

Both mellow sweet peppers and scorching chillies thrive in warm, sunny gardens. Hugely attractive with their fiery colours. Great candidates for containers outdoors or indoors.

Growing essentials

Sowing Sow seeds 1cm (½in) deep, 1cm (½in) apart, in pots under cover. Once seedlings are large enough to handle, transplant individually into 9cm (3½in) diameter pots and grow on under cover.

Planting When roots fill the pot, move into 30cm (12in) diameter containers or plant out once the risk of frost has passed. Space plants 40cm (16in) apart, supporting with stout canes.

Harvesting Harvest immature green fruits, or allow to colour up for sweeter or hotter flavours. If chillies are dried once mature, they can store for years.

Secrets of success

Set a propagator at 18–25°C (64–77°F) for good germination – especially for the hottest chillies.

Chillies are incredibly heat tolerant, growing well in hot summers.

Grow sweet peppers hydroponically under cover to significantly increase yields.

Feed large crops regularly with high potash organic liquid feeds, especially sweet peppers.

To ripen late summer chillies, reduce watering and partly unearth the plants.

Freeze gluts in sealed bags.

Aubergines
Solanum melongena

	J	F	M	A	M	J	J	A	S	O	N	D
SOW			●									
TRANSPLANT						●						
HARVEST								●	●			

FROM SOWING TO HARVEST
22–28 weeks

VARIETIES
'Violet Knight' F1 – productive with sausage-shaped, purple fruits

'White Knight' F1 – vigorous, yielding many white, sausage-shaped fruits

'Moneymaker No. 2' F1 – compact, abundant early purple fruits

'Raspberry Ripple' F1 – compact, striking purple fruits striped with white

ORIGINS
Warm and humid tropical Asia

ALSO GROWS IN
Zones 2 and 3 – if sufficiently warm

Zone 8 – grow compact varieties as houseplants

HARDINESS
Not hardy – killed by frosts

LIFE CYCLE
Perennial, grown as an annual

YIELD
★★★/☆☆☆☆

EASY TO GROW
6/10

From left to right: aubergines 'White Knight', 'Violet Knight', and 'Kermit'.

Popular in Asian and Mediterranean cuisine and enjoyed in a variety of dishes. Plants respond well to heat, yielding a significant number of fruits if strong root growth is encouraged.

Growing essentials

Sowing Sow seeds 5mm (¼in) deep, 10mm (½in) apart, in pots under cover. Once seedlings are large enough to handle, transplant into individual 9cm (3½in) diameter pots, grow on under cover.

Planting Keep young plants moist but not waterlogged, and temperatures at 16°C (60°F) or more. Aubergines perform best in 30cm (12in) pots. Pot up individually once well established, moving outside in June.

Harvesting Tie plants to a sturdy bamboo cane as they mature, keep well watered (but not waterlogged). Harvest fruits once full sized, before skins dull and wrinkle.

Secrets of success

Set a propagator at 20–24°C (68–75°F) for good germination. Cover seeds with a little compost.

Grafted plants are vigorous and quick to crop.

Grow in greenhouse borders for large plants with a huge harvest.

Give high-potash liquid feed when flowers appear, every 6–10 days.

Water mindfully – keep plants moist but not waterlogged or too dry.

Deter red spider mite by misting plants.

ZONE 1: SUNNY AND SHELTERED

Sweet potato
Ipomoea batatas

	J F M A M J J A S O N D
PLANT	
TRANSPLANT	
HARVEST	

FROM PLANTING TO HARVEST
18–24 weeks

VARIETIES

'Beauregard Improved' – red-skinned, orange-fleshed, cold tolerant

'Bonita' – cylindrical, cream skin, nutty, mild-tasting flesh

'Erato Violet' – elongated, deep purple, modest yield

'Evangeline' – sweet, deep orange, needs long summer

ORIGINS
Rich soils of tropical America, thriving in sunny areas

ALSO GROWS IN
Zone 3, if sheltered

HARDINESS
Not hardy – killed by frosts

LIFE CYCLE
Perennial, grown as an annual

YIELD
★★★/☆☆☆☆

EASY TO GROW
7/10

Breeding is fast making sweet potatoes a viable crop in temperate gardens. They'll prosper in a sheltered sunny corner, yielding sizeable sweetly-fleshed tubers, to enjoy as fries, wedges, jackets, and more.

Growing essentials

Starting Start plants off in spring as rooted cuttings, called "slips". Pot up individually into 20cm (8in) pots, grown on under cover at 18–20°C (64–68°F) until established.

Planting Harden off and transplant out 40cm (16in) apart, into soil enriched with garden compost and balanced granular feed. Plant through sheets of cardboard, water well, and cover with cloches if chilly.

Harvesting Harvest once foliage has yellowed and died down, but before harsh frosts. Carefully lift with a fork, and eat any that bruise immediately. Cure skins in a warm spot for a few days, then store somewhere warm.

Secrets of success

Create home-grown slips by sprouting healthy tubers in spring in a heated propagator.

Large tubers grow when sheets of cardboard prevent plants rooting in numerous places.

Train plants up poles or grow in containers where space is limited.

Ample water is essential. Water plants during dry spells to keep soil moist.

Eat young shoots and leaves from established plants as a green vegetable.

Flavours mellow in storage, often becoming sweeter with a fluffier texture.

Sweet potatoes produce a mass of foliage above ground, and delicious sweet tubers come late autumn.

Ten other star performers

Asparagus bean
Or yardlong bean, hinting at this tender annual's impressively lengthy pods. *Vigna unguiculata* subsp. *sesquipedalis* takes time to bulk up but then yields heavily. Dwarf and climbing forms are available. Eat them young.

Quinces
A hardy deciduous Mediterranean tree (*Cydonia oblonga*), grown for large, golden-yellow, pear-shaped fruits. Appreciates warmth, moisture, and free drainage. Highly aromatic, flavoursome, and excellent cooked, most varieties, like 'Vranja' and 'Meech's Prolific' are self-fertile.

Cinnamon vine
A hardy perennial climber (*Dioscorea polystachya*), growing freely on warm hills in China. Produces a mass of scrambling top growth in a warm spot and will tolerate some shade. Grown for sizeable edible tubers, and smaller aerial tubercles.

Chopsuey greens
A hardy annual, *Glebionis coronaria* syn. *Chrysanthemum coronarium* (or shungiku) grown for grey-green aromatic leaves. Quickly bulks up, producing attractive yellow flowers in summer. An excellent pollinator. Eat foliage young for best flavour.

Pineapple guava
Feijoa or *Acca sellowiana*. This borderline hardy evergreen shrub hails from South America, revels in sun, and tolerates wind. Growing to 2m (6½ft) with sausage-shaped green fruits and fleshy pink petals that melt in your mouth.

Chopsuey greens (left) and pineapple guava (right) enjoy bathing in sheltered sunshine.

Bay

Thriving in warmth, this southern European evergreen tree (*Laurus nobilis*) grows to 10m (33ft) tall, but is often clipped to keep smaller. It is wind tolerant, but winter chills can scorch leaves and split bark. The leaves have a beautifully aromatic warmth.

Achocha

Incredibly vigorous in warmth, this tender annual cucurbit (*Cyclanthera pedata*) quickly forms a sizeable plant yielding horn-shaped swollen fruits with a mild green-pepper flavour, from summer to autumn. Fruits of 'Bolivian Giant' are especially large.

Tomatillo

Also called Mexican husk tomato (*Physalis ixocarpa*), due to this fruit's origins and papery casing. Tender annual quickly growing from seed into lax, open plants, best staked. Yellow flowers and spherical green or purple fruits, excellent for salsas.

Kiwis

Scrambling, relatively hardy vines from mountainous East Asia. Kiwi fruit (*Actinidia deliciosa*) and kiwi berry (*A. arguta*) offer best fruiting. Self-fertile forms *A. d.* 'Jenny' and *A. a.* 'Issai' are easiest to accommodate, and highly productive.

Cape gooseberry

Another tender physalis (*P. peruviana*), native to Peru. Stake plants if exposed. Ripe orange fruits, known as "ground cherries", fall late summer in papery cases. They store, once ripe, for many weeks.

Bay (top left), cinnamon vine (top right), and tomatillo (bottom), will all grow vigorously in Zone 1.

Project:
Fan training fruit in the sun

Many fruit trees and bushes can be trained into two-dimensional shapes to bask on sunny walls or fences. Training isn't complex, you just need a basic understanding and the passage of time. Trained fruit is both highly ornamental, and hugely productive for the space it occupies.

Fan training can be applied to crops that bear fruit along the length of their stems; stone fruits such as peaches (pictured above), plums, and cherries, along with gooseberries and red, white, and pinkcurrants. Entire stems are fanned out in a symmetrical framework. At maturity older stems are periodically removed and new ones tied or trained into their place to keep the plant productive.

A fan-trained tree offers many benefits: primarily that the fruit captures as much sun as possible. This boosts sugar levels, great for sweet cherries, gages, apricots, nectarines, peaches, and figs. The two-dimensional shape makes barriers for pest and disease straightforward to use, fruit easier to harvest, and saves on groundspace. Pre-trained trees may yield faster, but they cost more, and a younger tree will bear a significant crop after three or four years.

ZONE 1: **SUNNY AND SHELTERED**

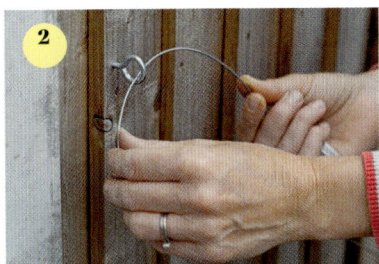
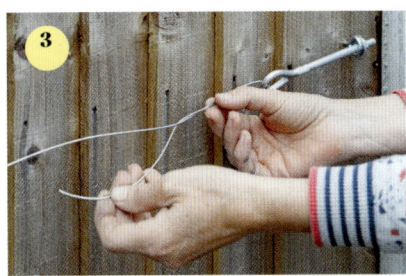

You will need

Tape measure
Drill
Screw-in vine eyes
10 or 12 gauge (8 or 10 AWG) galvanized wire
Straining eye bolts
Adjustable spanner
Jute twine
Bamboo canes

Steps

1 Drill pilot holes for pairs of vine eyes at 30cm (12in) height intervals, each pair spanning a 2–3m (6½–10ft) width. Insert eyes.

2 Attach 2–3m (6½–10ft) lengths of wire securely to each left eye.

3 Thread the wire through the straining bolt eye, secure the wire, pass the bolt thread through the right vine eye, and tighten the nut using an adjustable spanner to tension the wire.

4 Choose a young plant with strong limbs growing left and right, plant, firm gently, water. Using twine, tie bamboo canes firmly to the wire, where the limbs will ultimately grow in a symmetrical fan pattern.

5 Tie young limbs to the canes – make knots loose enough that stems have space to grow.

6 Cut out branches growing into or away from the wall or fence.

7 As the tree matures, tie new stems to canes, and secure canes to wires. Thin out old sideshoots of these stems as they mature, and tie in new ones.

You can quash any concerns that such seemingly inhospitable areas won't support your edibles – many crops from maritime, arid, and mountainous areas were built not just to survive, but to thrive, in these environments. The botanical adaptations are multiple. Asparagus, salsify, and seakale have fleshy roots that delve deep into sandy soils; globe artichokes and goosefoots possess waxy foliage packed with fine hairs to reflect away strong sunlight; and agretti and purslane boast plump succulent leaves perfectly designed to hold onto any moisture available. When nature throws us such extremes, simply pair them with these edibles.

Zone 2
Sunny, open, and dry

Asparagus
Asparagus officinalis

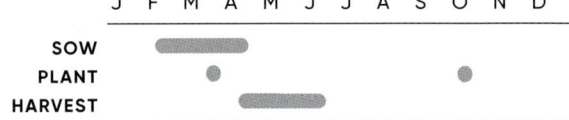

FROM PLANTING TO HARVEST
2–3 years

VARIETIES
'Gijnlim' – all-male F1 hybrid, fast growing, thick green spears

'Guelph Millenium' – all-male F1 hybrid, vigorous, cold tolerant

'Pacific Purple' – tender, sweet, purple spears

'Vittorio' – robust all-male F1 for "forced" white spears

ORIGINS
Southwest Europe and Mediterranean shorelines

ALSO GROWS IN
Zone 1 and Zone 3 – if not waterlogged

HARDINESS
Hardy

LIFE CYCLE
Perennial

YIELD
★★/☆☆☆☆☆

EASY TO GROW
8/10

Asparagus has luxury status due to its short harvest season, space-hungry nature, and unique flavour. It takes patience to establish but then, expect fresh, home-grown spears for a decade or longer. Eat fresh for the sweetest flavour – steamed, grilled, or boiled and seasoned with butter.

Growing essentials

Sowing Sow individually in 9cm (3½in) pots in spring. Or start off as 1-year-old crowns (dormant roots). Place 40cm (16in) apart in a trench 10cm (4in) deep, cover with soil. Water well the first year.

Planting plant out early summer. Harvest crowns from their second year, seed-raised plants from their third.

Harvesting Harvest for 8 weeks the first year, and 10 thereafter. Use a sharp knife or secateurs to cut spears 1–2cm (½–¾in) below soil level, once 20–25cm (8–10in) high.

Secrets of success

Sowing seed is inexpensive. Set your propagator to 16–20°C (60–68°F).

Plant into weed-free beds – ensure no perennial weeds abound.

Nurture young plants to establish. They are very drought resistant once mature.

Sprinkle balanced granular feed such as blood, fish, and bone onto beds after the harvest.

Asparagus is happy in free-draining soils, where mature plants show strong drought resistance.

Stop cutting in summer to allow foliage to develop so that crowns can recover – support with stakes and twine.

Remove old fern in November, do not compost if any asparagus beetle (see page 206) is present.

ZONE 2: SUNNY, OPEN, AND DRY

Salsify and scorzonera

Tragopogon and *Scorzonera* species

	J F M A M J J A S O N D
SOW	▬▬ ▬▬
TRANSPLANT	Do not transplant (sow direct)
HARVEST	▬▬▬

FROM SOWING TO HARVEST
22–30 weeks

VARIETIES

Salsify 'Sandwich Island Mammoth' – thick, white, vigorous

Salsify 'Scorzobianca' – slender, white skin and flesh

Scorzonera 'Russian Giant' – vigorous black roots, cream flesh

Scorzonera 'Long Black Maxima' – long, black skin, white flesh

ORIGINS
The coasts and dunes of southern Europe

ALSO GROWS IN
Zones 1, 3, and 4

HARDINESS
Fully hardy

LIFE CYCLE
Biennial (salsify), perennial (scorzonera), both grown as an annual

YIELD
★★★/☆☆☆☆

EASY TO GROW
8/10

Incredibly drought resistant, these crops are rarely available in shops. The slender roots grow in abundance and can be seasoned, sauteed, or roasted then peeled as a gourmet side. Delicate and sweet – some even say they are reminiscent of oysters.

Growing essentials

Sowing Sow seeds in rows 30cm (12in) apart, 3cm (1¼in) deep, directly into well-dug soil. Water if soil is dry. Once seedlings emerge thin out to 10cm (4in) apart. Keep rows weeded until seedlings establish – afterwards plants are low maintenance. Remove larger weeds.

Harvesting Begin harvesting in early autumn. On open soils roots penetrate deeply so use a strong fork. Lift as and when required; roots store best in the ground.

Secrets of success

Seed viability is short, buy fresh packets every year or two.

Don't attempt to transplant – the long taproots resent disturbance.

Boost yields by watering young plants in prolonged periods of drought.

Mulch roots with straw if harsh frost is forecast, to make harvesting easier.

Blanch young leaves with an upturned pot as a tasty extra crop.

Flowers attract wildlife – leave overwintered plants to grow if you have room.

The thick, long roots of salsify give plants excellent drought resistance. Top growth can be blanched and eaten, too.

Summer purslane
Portulaca oleracea

	J	F	M	A	M	J	J	A	S	O	N	D
SOW					▬	▬	▬					
TRANSPLANT						▬	▬					
HARVEST							▬	▬	▬	▬		

FROM SOWING TO HARVEST
12–16 weeks

VARIETIES
Usually sold as the upright-growing species, small-leaved creeping forms also found

ORIGINS
Tropical regions across the world, with many variations in form

ALSO GROWS IN
Zones 1, 3, and 5

HARDINESS
Not fully hardy – killed by frosts

LIFE CYCLE
Short-lived perennial, grown as annual

YIELD
★★★★½/☆☆☆☆

EASY TO GROW
9/10

An adaptable survivor. Its waxy leaf cuticle, succulent foliage, and ability to function in arid conditions cements it as a crop for our climate's future. Luckily, it's also very productive, easy to cultivate, and refreshingly delicious.

Growing essentials

Sowing Sow the tiny seeds thinly into small modules of moist seed compost and place in a propagator at 18–20°C (64–68°F). Pot up once large enough.

Alternatively sow outside in June, in drills 30cm (12in) apart. Thin seedlings to 20cm (8in) apart.

Planting Plant potted purslane outside when all risk of frost has passed. Once established and if left to flower, summer purslane will freely self-seed.

Harvesting Harvest once plants have bulked up sufficiently. Repeatedly pinch out and eat growing points – this encourages stocky, robust plants with multiple sideshoots.

Secrets of success

Do not cover seeds with compost, as sunlight is required for germination.

If allowed to set seed in your garden, you generally won't need to sow summer purslane under cover again.

Self-sown seedlings show far more resilience to drought than transplanted pots of purslane.

Regularly pinch out in summer. As autumn draws close, plants will naturally flower.

Prostrate small-leaved forms of summer purslane (occasionally sold as *P. sativa* or *P. retusa*) make excellent groundcover in drought-prone areas.

Eat harvests fresh, or wilt down in seasoned butter.

You will come across both upright (top) and prostrate (bottom) forms of summer purslane – both have excellent drought tolerance.

ZONE 2: SUNNY, OPEN, AND DRY

Lima beans
Phaseolus lunatus

	J F M A M J J A S O N D
SOW	▬▬
TRANSPLANT	▬▬
HARVEST	▬▬▬▬▬▬

FROM SOWING TO HARVEST
16–20 weeks

VARIETIES
Named varieties (including dwarf types) occasionally available, seeds are usually sold as "lima beans" or "butter beans"

ORIGINS
Hot and dry South America, naturally more heat and drought tolerant than other beans

ALSO GROWS IN
Zone 1 and also Zone 3 if near a windbreak

HARDINESS
Not hardy – killed by frosts

LIFE CYCLE
Perennial, grown as annual

YIELD
★★★/☆☆☆☆

EASY TO GROW
7/10

Also known as butter bean or sieva bean, seeds of this tropical legume, once boiled then slow-cooked, offer a protein- and fibre-rich ingredient for casseroles, soups, paellas, and salads. Highly productive in long hot summers and dried gluts store incredibly well.

Growing essentials

Sowing Sow seeds 3cm (1¼in) deep, in individual pots under cover in spring. Once seedlings are fully emerged, harden off. Alternatively, sow outdoors once the risk of frost has passed.

Planting Plant out alongside beanpoles set at 40cm (16in) apart, or in rows 40cm (16in) apart. Protect young plants from strong winds, water well until established. Climbing forms will twine up supports unaided.

Harvesting Attractive flowers appear in summer, followed by pods, ultimately filled with plump beans. Harvest once fully mature. Dry beans for storage.

Secrets of success

Lima beans are closely related to runner beans – treat identically.

Avoid sowing into cold or waterlogged conditions – wait until full spring.

Erect sturdy supports for climbing varieties. Grow dwarf forms in pots.

Add ample organic matter to beds, for strong root systems.

Pods form in dry weather but may be more fibrous than if produced in wetter seasons.

Dry then rehydrate seeds for a protein-rich food. Boil rapidly for 20 mins, then simmer until tender.

Hailing from arid regions, lima beans are a good choice for drier plots, freely producing large beans.

ZONE 2: SUNNY, OPEN, AND DRY

Figs
Ficus carica

	J F M A M J J A S O N D
PLANT	▬▬▬▬▬ ▬▬▬
HARVEST	▬

FROM PLANTING TO HARVEST
1–4 years

VARIETIES
'Brown Turkey' – reliable, productive, delicious fully ripe

'Rouge de Bordeaux' – deep purple fruits, exquisite flavour

'Panachée' – reliable, vigorous, vivid green- and cream-striped

'Babits' – very cold-hardy Hungarian, deep purple, richly flavoured

ORIGINS
Asia and Southern Europe, revelling in sunshine

ALSO GROWS IN
Zones 1 and 3

HARDINESS
Hardy

LIFE CYCLE
Perennial

YIELD
★★★/☆☆☆☆

EASY TO GROW
8/10

One of the most delicious gardening experiences has to be eating a home-grown fig. The quality of the fruit – and the ability to harvest it at perfect ripeness – entitle you to supreme joy. The trees are highly ornamental, too, and will crop generously once mature.

'Brown Turkey' is a reliable and heavy cropping variety in temperate gardens. Pick figs when they are perfectly ripe.

Growing essentials

Planting Plant in a sunny position in well-drained soil dug over with garden compost or well-rotted manure.

Pruning Fig trees can be left unpruned, but work brilliantly as fans against walls or fences (see pages 74–75). Build up a framework of branches with initial training. Thin these out annually, in late spring, once mature.

Harvesting Harvest when fruits are soft and heavy with juice for best flavours. Eat quickly, before they spoil.

Secrets of success

Protect overwintering figs from frost – moving under cover or insulating with fleece and straw.

Growing in pots in a tunnel or greenhouse will encourage all fruit to reach maturity.

Restrict the size of trees in the ground by lining planting pit sides with paving slabs.

Store gluts via freezing for cooking later, by preserving as jam, or by drying.

Pinch out growing tips in June to encourage good numbers of overwintering figlets.

Propagate via hardwood cuttings or low stems rooting into soil.

ZONE 2: SUNNY, OPEN, AND DRY

Agretti
Salsola soda

	J F M A M J J A S O N D
SOW	▬▬
TRANSPLANT	▬
HARVEST	▬▬▬▬▬

FROM SOWING TO HARVEST
16–20 weeks

VARIETIES
Seeds sold as basic species **Salsola soda**. Also look for the closely related *S. komarovii* (saltwort)

ORIGINS
The Mediterranean Basin, in free-draining, maritime conditions

ALSO GROWS IN
Zone 1 and Zone 3 – if not waterlogged

HARDINESS
Borderline – killed by harsh frosts

LIFE CYCLE
Annual

YIELD
★★★★/☆☆☆☆

EASY TO GROW
8/10

A delicacy in Italy and Greece, with other regions now catching on to how useful agretti is. Very drought and heat tolerant once established, with the potential to improve the quality of saline soils. Leaves are succulent and plentiful, delicious both raw and cooked.

Growing essentials

Sowing Sow the corky seeds 1cm (½in) deep in pots, under cover in spring. Once seedlings are large enough, transplant individually into small pots, and grow on under cover until all risk of frost has passed.

Planting Plant in a sunny, well-drained bed in late May or early June. If grown outside last year, self-sown seedlings may appear in spring – these will be more drought tolerant than transplants.

Harvesting Allow plants to bulk up, watering those raised in pots until well established. Harvest by pinching out the growing points regularly.

Secrets of success

Obtain fresh seed or save your own from last year's plants for good germination.

Grow in unheated glasshouses or polytunnels to maturity as a year-round crop – plants can withstand a light frost.

Sow direct into beds once risk of harsh frosts has passed for very drought-resilient plants.

The fleshy, needle-like leaves of agretti allow it to grow well in dry soils.

Guarantee tender growth by regularly pinching out plants' growing points to encourage vigorous side-shooting.

Collect seeds once fully mature and brown. Store and sow in spring.

Delicious simply wilted down in olive oil, then seasoned with lemon juice.

Goosefoots
Chenopodium species

	J F M A M J J A S O N D
SOW	— — —
HARVEST	— — — —

FROM SOWING TO HARVEST
10–20 weeks

VARIETIES
C. album – fat hen. Abundance of grey-green waxy leaves

C. berlandieri – huauzontle. Broccoli-like seed heads

C. bonus-henricus – good King Henry. Perennial, grown for leaves

C. giganteum – tree spinach. Striking hot pink shoots

ORIGINS
Tropical and temperate climates

ALSO GROWS IN
Zones 1 and 3

HARDINESS
Not fully hardy – killed by harsh frosts

LIFE CYCLE
Mostly annuals, all treated as such

YIELD
★★★★★/☆☆☆☆☆

EASY TO GROW
9/10

Huauzontle (left), good King Henry (centre), and tree spinach (right) all offer leafy pickings in drought-prone spots.

The goosefoot family bring us an abundance of tender, vitamin-packed leaves, with many grown for their protein-rich seeds. All opportunistic self-seeders that will grow vigorously in the right conditions. They're crops that many will deem essential, due to their drought tolerance.

Growing essentials

Sowing Sow seeds in large modules under cover, or directly in soil for optimum performance. Once risk of frost has passed, rake over a bed, then sprinkle seeds over surface, in rows 30–40cm (12–16in) apart.

Aftercare Thin seedlings to 20–30cm (8–12in) apart. Water in dry spells so plants can establish. Little water is needed in summer, but giving more can boost yields.

Harvesting Harvest by pinching out shoot tips of fat hen, good King Henry, and tree spinach once plants have bulked up.

Harvest huauzontle seed while still green. Soak in water overnight, then cook them.

Secrets of success

To sow under cover, set propagators at 16–18°C (60–64°F). This can give earlier harvests for baby leaf crops.

Goosefoots resent root disturbance – make sure to plant out before any root congestion occurs.

Harvest regularly for a succession of tender, soft sideshoots.

Eat cooked, rather than raw if consuming large quantities, to reduce saponin and oxalic acid levels.

Save seed by allowing one or two plants to run to seed.

Cover seed heads in paper or muslin bags, so seeds don't fall to the soil and take over as weeds.

ZONE 2: SUNNY, OPEN, AND DRY

Grapes
Vitis species

	J F M A M J J A S O N D
PLANT	▬▬▬▬ ▬▬
HARVEST	▬▬

FROM PLANTING TO HARVEST
1–3 years

VARIETIES
'Fragola' – dessert type, abundant, strawberry-flavoured, pink, seeded

'Siegerrebe' – golden seeded, mild muscat flavour, dual purpose

'Boskoop Glory' – dual-purpose black grape, few pips, mildew resistant

'Chardonnay' – Most popular UK vineyard grape, white, seeded

ORIGINS
Europe, the Middle East, and southwest Asia on warm, sunny slopes, in free-draining soil

ALSO GROWS IN
Zones 1, Zone 3, and Zone 4 – choose wine grapes in cooler zones

HARDINESS
Fully hardy

LIFE CYCLE
Perennial

YIELD
★★★★/☆☆☆☆

EASY TO GROW
7/10

Evoking sun-drenched holidays and sociable evenings, the grapevine is one of those edibles that's desirable on every plot. Whether you choose an eating or winemaking grape, their robust, trouble-free, and generous nature will provide years of harvests.

Growing essentials

Planting Plant dormant vines into free-draining, well-prepared soil. Feature a single plant on a pergola or similar structure, or plant multiples in rows.

Pruning Best pruned in early winter, after leaf fall, for a good stem framework. Cut back shoots in summer to control growth.

Harvesting Harvest bunches in autumn, once sugar levels peak. Remove with secateurs and eat fresh, or press for juice or wine-making.

Secrets of success

Grapes are happy exposed. Remember, sun is essential to boost sugar levels.

Young vine leaves are edible, and popular stuffed with savoury fillings.

Thin dessert grape bunch numbers to boost fruit size (fruitlets can also be thinned).

Train garden grapes over tall structures. Wine grapes are kept on a low framework.

Mildew can spoil eating quality of dessert grapes – look for resistant varieties.

Autumn leaf colour can be stunning – prune once faded to deter sap bleeding.

Ample sunshine and warmth, plus fertile, free-draining soils help to boost ripening and sugar levels of grapes.

Globe artichoke

Cynara cardunculus Scolymus Group

	J	F	M	A	M	J	J	A	S	O	N	D
SOW			▬	▬								
TRANSPLANT						▬						
HARVEST							▬	▬				

FROM SOWING TO HARVEST
68–72 weeks

VARIETIES

'Green Globe' – widely grown, large spherical green buds

'Purple de Provence' – ornamental, less hardy, purple-tinged globes

'Gros Vert de Lâon' – vigorous, heritage, large green hearts

C. cardunculus – cardoon. Edible blanched stems

ORIGINS
The Mediterranean, in sunny, hot, free-draining soil

ALSO GROWS IN
Zone 1 and Zone 3

HARDINESS
Hardy, but damaged in severe winters

LIFE CYCLE
Perennial

YIELD
★★/☆☆☆☆☆

EASY TO GROW
8/10

Handsome, huge, and architectural, the globe artichoke is one thistle that we welcome to the plot. The steely grey leaves are topped with fat, ball-shaped buds in early summer that are made for butter and Hollandaise. Blanched cardoon stems are equally decadent.

Growing essentials

Sowing Sow seeds singly, into 9cm (3½in) diameter pots under cover in spring. Grow on until large enough to plant out. Wait until risk of frost has passed, then harden off.

Planting Enrich soil with well-rotted manure or organic matter, giving each plant 1m (3ft) square spacings. Water until established. Mature plants need little care.

Harvesting Harvest buds once full sized, before scales open. Smaller sideshoot buds appear after the terminal ones. Blanch mature cardoon stems for 3 weeks in cardboard tubes.

Secrets of success

Seed-raised plants take over a year to bud – buy plants for speedier results.

Protect new shoots from slugs – especially if blanching cardoons.

Bees love the flowers – large, purple, and thistle-like, appearing in midsummer.

Cut down woody stems in late summer to encourage fresh leaves.

Divide every few years, in autumn or spring, to keep productive.

Mulching in winter with bark protects plants from chills.

'Green Globe' artichoke yields delicious harvests on drought-proof plants.

ZONE 2: SUNNY, OPEN, AND DRY

Amaranth/ Quinoa

Amaranthus species

	J	F	M	A	M	J	J	A	S	O	N	D
SOW				▬	▬							
TRANSPLANT						▬						
HARVEST							▬	▬	▬	▬		

FROM SOWING TO HARVEST
2 weeks (leaf), 24 weeks (grain)

VARIETIES

'Red Army' – striking, abundant ruby leaves and flower spikes

'Hot Biscuits' – traditional green for callaloo, brown flower spikes

'Dreadlocks' – attractive, lush leaves, purple ball-like flowers

ORIGINS
Caribbean and central American tropics – quinoa was a staple Aztec grain

ALSO GROWS IN
Zone 1 and Zone 3 – if sufficiently warm

HARDINESS
Not hardy – killed by frosts

LIFE CYCLE
Annual

YIELD
★★★★/☆☆☆☆☆

EASY TO GROW
8/10

Known as callaloo in Caribbean cooking, this quick-growing, leafy crop revels in heat. It's very attractive, coming in several vivid colours. Leave to flower for the protein-rich quinoa seeds.

Growing essentials

Sowing Sow undercover but don't cover seeds. Move seedlings into individual pots once large enough. Seed can be sown in soil, once risk of frost has passed. Space drills 30cm (12in) apart, and thin seedlings to 20cm (8in) apart.

Water during drought until established – mature plants will survive dry periods.

Harvesting Pinch out growing tips and sideshoots to harvest as a leafy vegetable. For grain, stop pinching out early July and then add supports if they become top-heavy.

Secrets of success

For best indoor germination maintain temperatures of 18–22°C (64–71°F).

Direct sowings are more resilient to drought than transplants.

Eat thinnings and spare seedlings. Amaranth makes a nutrient-dense sprouting seed.

Pinch out regularly for leaf crop, otherwise plants self-sow freely.

Amaranth (here, 'Red Army') can be harvested as a leafy crop, or allowed to mature and produce grains.

Harvest whole spikes when seeds start falling from flowers, then dry on a sheet before threshing.

To remove saponins, soak quinoa seeds and rinse until froth disappears.

Ten other star performers

Red orach
Highly ornamental red selection (*Atriplex hortensis* var. *rubra*) of borderline hardy European annual. Green forms also found. Self-seeds freely, best sown direct rather than transplanted. Pinch tips to harvest and allow to set seed for collection.

Garlic
Allium sativum. Plant in autumn, 12cm (4¾in) deep in free-draining soil, for drought resilience. Spring-planting gives smaller yields but is useful for "scapes" and "rounds" – small, single bulbs. Early-maturing 'Rhapsody Wight' is my favourite.

Oregano
Flavoursome southern Mediterranean herb, excellent drought resistance, particularly varieties with smaller, hairier leaves like *Origanum majorana* and *O. dictamnus*. One of the most versatile culinary herbs, along with thyme. Dries very well for storage.

Oca
Pretty little South American plant (*Oxalis tuberosa*). Borderline hardy, producing masses of knobbly tubers below shamrock-style leaves. Lift just before first frosts for maximum yield. It also makes an excellent and attractive groundcover plant.

New Zealand spinach
Borderline hardy perennial from Oceania, grown frequently as an annual. Quick to grow and found on dunes and shorelines, forming

The flowers of oregano (left) attract pollinators. Red orache (right) makes a colourful salad addition.

Seakale plants (left) can be forced in spring. Opium poppies (centre) produce abundant seeds and oca (right) bulks up freely for autumn harvests.

a groundcover of succulent arrow-shaped leaves. Keep pinching out tips to maintain tender, palatable harvests.

Jerusalem artichoke

Helianthus tuberosus. American hardy perennial and excellent shelterbelt plant, to 2m (6½ft) tall. Clusters of knobbly tubers late autumn. Store underground until needed. Some varieties like 'Fuseau' flower freely, making them excellent for wildlife.

Seakale

Hardy perennial (*Crambe maritima*) native to European shorelines. Supremely drought resistant. Highly ornamental, with glaucous leaves and huge frothy flower spikes. Force shoots in spring under large pots, removing in early summer so that plants can recover.

Opium poppy

Hardy annual (*Papaver somniferum*) making excellent protein-rich seeds produced in abundance. Drought resistance is maximized if plants are allowed to self-seed. Sow direct the first year. Flowers highly decorative and useful for pollinators.

Horn of plenty

Fedia cornucopiae, a hardy annual salad also known as African valerian, grows to 30cm (12in) tall. Great heat and drought tolerance. Harvest young shoots, but allow some to mature, for pink flowerheads and seeds for collection.

Turkish rocket

Hardy perennial from the cabbage family. Known as hill mustard or *Bunias orientalis*. Enjoys full sun and becomes more drought tolerant with age. Eat young leaves as a tender green, and unopened flower spikes like broccoli florets.

Project: Plant supports for windy sites

An exposed location needn't limit what you grow. Choose crops with "windproof" growth habits, and accommodate taller fruit trees with a sturdy support system – handy for windbreak plants yet to mature, or when allotment tenancy agreements dictate that only edible species are allowed to be planted.

With careful choice, there are numerous trees bearing edible crops that can tolerate an open, windy position. Staking taller specimens for their initial years of establishment allows them to gain a firm foothold, and ultimately your efforts will be rewarded with strong, robust trees that support good harvests. The low staking system shown here is ideal for many of the edibles listed below, if grown as a single-stemmed plant (multi-stemmed plants would be less top heavy). Applying a thick organic mulch after planting will minimize any soil moisture losses.

Fruit trees for windy sites:

Damsons (*Prunus domestica* subsp. *insititia*)
Sloes (*Prunus spinosa*)
Plums (*Prunus domestica*)
Elderberries (*Sambucus nigra*)
Crab apples (various *Malus* species)
Serviceberry (*Amelanchier* species)
Sea buckthorn *(Hippophae rhamnoides)*

Fruit canes and bushes for windy sites:

Blackberries (*Rubus*)
Hybrid berries (combinations of various species)
Blackcurrants (*Ribes nigrum*)
Gooseberries (*Ribes uva-crispa*)

You will need

Spade
Two tree stakes
Mallet or post rammer
Timber crossbar
Tree tie
Drills and screws

Steps

1 Plant your chosen crop between autumn and spring, when soil is neither frozen nor waterlogged. If planting a rootballed or potted tree, gently tease out any spiralling roots; fan out the roots of bare-root trees.

2 Work soil between the roots, firm gently, check soil level, water, and mulch.

3 Insert two tree stakes, one either side of the tree, driving them at least 40cm (16in) into the soil using a mallet or post rammer.

4 Securely screw the crossbar to the top of the two stakes so that the trunk rests against it.

5 Weave a tree tie around the trunk.

6 Screw each end of the tie to the crossbar. As the trunk thickens, loosen the tie.

tip

If you want to plant a row of canes or bushes, choose those listed (left), and modify the wire system shown on pages 74–75, inserting tree stakes as the vertical supports.

With ample sunshine, warmth, and moisture, abundant growth is guaranteed. This zone is ideal for those "heavyweight" edibles that bulk up rapidly as soon as temperatures allow. Courgettes, winter squashes, melons, and cucumbers will grow before your eyes above ground. Slide your fork prongs into the soil come the end of the growing season and potatoes, mooli radish, and yacon will have bulked up beautifully for your winter stores, along with dahlias and cannas. Podded crops, too, such as French and lablab beans will yield generously. Essentially, brace yourself in such conditions for plentiful pickings.

Zone 3
Sunny and moist

ZONE 3: SUNNY AND MOIST

Winter squashes and pumpkins
Cucurbita species

```
       J F M A M J J A S O N D
SOW              ▬▬▬
TRANSPLANT          ▬▬
HARVEST                   ▬▬▬
```

FROM SOWING TO HARVEST
22–26 weeks

VARIETIES

'Crown Prince' F1 – large, 25cm (10in) diameter squashes. Blue skin, orange flesh

'Hunter' F1 – many 20–40cm (8–16in) long butternuts, rich orange flesh

'Rouge Vif D'Etampes' – large 35–45cm (14–18in) diameter, rich orange pumpkins

'Uchiki Kuri' – many small 18–23cm (7–9in) diameter squashes. Orange skin and flesh

ORIGINS
Domesticated in Central America, grown extensively in tropical and temperate climates

ALSO GROWS IN
Zone 1

HARDINESS
Not hardy – killed by frosts

LIFE CYCLE
Annual

YIELD
★★★★½/☆☆☆☆

EASY TO GROW
9/10

A hugely important plant family. Yields are sizeable, when hungry and thirsty appetites are respected. The nutty flesh is rich and delicious. There are ways to curb their vigour, if space is limited.

Growing essentials

Sowing Sow seeds 3–4cm (1¼–1½in) deep, singly, in pots or large modules, under cover in spring. Harden off once large enough. Sow seeds directly once soil is warm.

Planting Dig plenty of well-rotted organic matter into bed, then plant squashes 1m (3ft) apart. Water well into the next few weeks. Redirect meandering stems towards the bed.

Harvesting Harvest squashes and pumpkins once mature and well coloured. Remove with a short length of stalk attached. Store somewhere frost-free.

Secrets of success

Keep soil moist by sinking upturned bottles alongside plants to act as a water reservoir. This also helps deter powdery mildew (see page 212).

'Uchiki Kuri' squash (top) and butternut 'Hunter' (bottom) bulk up beautifully with warmth and moisture.

Train stems up tripods or arches to restrict spread.

Plant through sheets of cardboard for low-maintenance beds.

Compact varieties include butternut 'Butterbush', pumpkin 'Amazonka', and squash 'Table King'.

To cure skins for winter, remove shading leaves and store fruits in sunshine.

Steam young tips, stuff and shallow fry the blooms.

ZONE 3: SUNNY AND MOIST

French and drying beans

Phaseolus vulgaris

	J F M A M J J A S O N D
SOW	▬▬
TRANSPLANT	▬▬
HARVEST	▬▬▬▬

FROM SOWING TO HARVEST
16–20 weeks

VARIETIES

'Cobra' – productive climber, numerous mid-green pods

'Delinel' – dwarf bean, high quality, slender green pods

'Lingua di Fuoco' – climbing borlotto, for drying. White and red pods

'Sonesta' – productive yellow-podded dwarf. Buttery, delicious

ORIGINS
America, now grown throughout warmer regions of the globe

ALSO GROWS IN
Zone 1

HARDINESS
Not hardy – killed by frosts

LIFE CYCLE
Annual

YIELD
★★★★/☆☆☆☆☆

EASY TO GROW
7/10

A staple world-wide, thanks to its versatility and heavy yields. Succulent, juicy, and sweet pods, freshly picked – a range of colours makes a beautiful plate. Left to dry, the beans transform into a hearty harvest.

Growing essentials

Sowing Sow seeds 2–3cm (¾–1¼in) deep, individually, in 9cm (3½in) pots under cover in spring. Grow on under cover until all risk of frost has passed, then gradually harden off.

Planting Transplant into a sunny spot, adding organic matter to the area. Space dwarf types 30cm (12in) apart in rows. Place climbers 20cm (8in) apart in a circle or double row, one plant per 2.5m (8ft) cane/pole.

Harvesting Begin harvesting pods when large enough. Gently pull away, and harvest every few days for a succession of beans.

Secrets of success

Dwarf beans crop in flushes, climbers yield steadily all summer.

Cold, wet soils encourage root rot. Pre-warm soil in spring with cloches.

Mulch new plantings to retain soil moisture, and water well until established.

Sow a later batch of dwarf types into the ground, to crop once early plants fade.

If growing for drying, harvest individual pods once fully swollen.

Soak dried beans overnight, boil rapidly for 20 minutes then simmer until tender.

Zone 3 supports ample pods of dwarf beans 'Polka' (left) and 'Yin Yang' (right).

Asian broccolis

Brassica species

	J F M A M J J A S O N D
SOW	▬▬
TRANSPLANT	▬▬▬
HARVEST	▬▬▬▬▬

TIME FROM SOWING TO HARVEST
6–10 weeks

VARIETIES
Choy sum – large, paddle-shaped, shiny green leaves, yellow flowerheads

Chinese broccoli – or *kai-lan*, glaucous large leaves, white flowerheads

ORIGINS
Warm and moist regions of Asia

ALSO GROWS IN
Zones 1 and 5

HARDINESS
Not hardy – killed by frosts

LIFE CYCLE
Annual

YIELD
★★★/☆☆☆☆

EASY TO GROW
7/10

These quick-to-crop plants are a great shortcut for broccoli lovers, maturing in less than two months in optimum conditions. Moisture is essential for the production of the lush leaves, and pickings allow cooks to create the most tender and delicate of dishes.

Growing essentials

Sowing Sow for an early crop, under cover in modules. Sow 1–2 seeds per cell, and transplant outside 30cm (12in) apart, under cloches early to mid-spring. Protect from slugs and snails.

As soil conditions warm in spring, sow directly into soil. Excavate a drill 5cm (2in) deep, and sow seeds 1–2cm (½–¾in) apart. Thin (and eat) emerging seedlings to 30cm (12in) apart.

Harvesting Harvest a few leaves when plants are large enough – leave sufficient foliage so they can re-sprout. The broccoli-like flowering shoots are best harvested just before flowers open.

Secrets of success

These crops enjoy cool weather – sow late summer for autumn harvests.

Warm, moist conditions may encourage slugs – be vigilant with controls.

Applying an organic mulch keeps roots moist and encourages the best yields.

Avoid drought stress, especially during prolonged hot spells in summer.

Plants can be sheared off as a cut-and-come-again leaf crop, or left to produce flower shoots.

The closely related broccoli raab, or rapini, is also worth growing in Zone 3.

Both choy sum (left) and Chinese broccoli (right) grow rapidly in sunny, moist conditions.

ZONE 3: SUNNY AND MOIST

Courgettes and summer squashes
Cucurbita pepo

	J F M A M J J A S O N D
SOW	▬▬
TRANSPLANT	▬▬
HARVEST	▬▬▬▬

FROM SOWING TO HARVEST
12–16 weeks

VARIETIES
'Defender' F1 – productive deep green-skinned courgette

'Coucourzelle' – Italian courgette, many fruits, striped mid- and deep green

'Golden Dawn' F1 – high-yielding yellow courgette

'Sunburst' F1 – attractive yellow-fruited pattypan-style summer squash

ORIGINS
Courgettes were developed in South America from wild squashes, they revel in warm, moist, fertile soils

ALSO GROWS IN
Zone 1

HARDINESS
Not hardy – killed by frosts

LIFE CYCLE
Annual

YIELD
★★★★½/☆☆☆☆

EASY TO GROW
8/10

The flavour of freshly harvested courgettes and summer squashes is incomparable. Super-sweet and nutty, just one or two plants keeps you well-stocked all summer long.

Growing essentials

Sowing Sow seeds, singly, in 9cm (3½in) pots under cover in spring. Grow on, with plenty of space and warmth. Harden off once risk of frost has passed.

Planting Enrich soil with ample garden compost and general-purpose granular feed. Plant courgettes and squashes 1m (3ft) apart, water well, and irrigate during dry spells.

Harvesting Harvest fruits when large enough. Use a sharp knife to avoid damaging stems or plants. For best flavour consume as soon after picking as possible.

Secrets of success

Sow seeds outside, in early summer, if you struggle for undercover propagation.

Sink an upturned bottle alongside each plant to apply a steady, generous volume of water.

Both courgettes and summer squashes are available in a wide range of shapes and colours. Mature plants yield prolifically, especially after summer rainfall.

For space-limited gardens purchase climbing forms like 'Shooting Star' or compact ones like 'Piccolo'.

Flowers and young shoots can be harvested and eaten from established plants.

Harvest while small for the best flavours and more fruits.

If powdery mildew is a problem look for varieties with resistance (e.g. 'Defender').

Broccoli

Brassica oleracea var. *italica*

	J F M A M J J A S O N D
SOW	▬▬▬
TRANSPLANT	▬▬
HARVEST	▬▬▬▬

FROM SOWING TO HARVEST
14–18 weeks

VARIETIES
'Ironman' F1 – vigorous, high-yielding, mid-green heads

'Purple Rain' F1 – striking and ornamental, rich purple harvests

'Inspiration' F1 – tenderstem selection, heads on tall shoots

ORIGINS
Bred in Italy, diversified in America and Asia

ALSO GROWS IN
Zone 1, Zone 4, and Zone 5

HARDINESS
Not fully hardy – killed by harsh frosts

LIFE CYCLE
Annual

YIELD
★★★★/☆☆☆☆

EASY TO GROW
7/10

Broccoli, or calabrese, is a hugely popular crop. Quick to mature and productive, if grown in ideal conditions. Its mild flavour makes it popular among children, and tenderstem selections give it gourmet status.

Growing essentials

Sowing Sow seeds 1cm (½in) deep in modules, under cover in spring for early crop. Alternatively sow seeds directly into soil once sufficiently warm. Sow two seeds per station, thinning later to one.

Planting Harden off and transplant outside after risk of frost. Plant in rows 40cm (16in) apart, with plants 30cm (12in) apart. Water well and remove weeds until established.

Harvesting Harvest heads once large enough, while buds are still tightly closed. Smaller sideshoots will ultimately appear, and these can be cut as a second harvest.

Secrets of success

- **Sow two seeds per cell** or station, thinning once germinated to leave the strongest.
- **Sow in February** for an earlier crop, plant out under cloches in Zone 1.
- **Net against butterflies and birds** – use a 5mm (¼in) mesh held away from plants.
- **Stalks make a tasty harvest** – peel and eat raw when the main head is cut.
- **Achieve crops in pots** – plant three plants per 40cm (16in) tub. Yields are smaller, but just as delicious.
- **Gluts freeze well** – calabrese frequently matures in unison, blanch then bag up.

Broccoli plants will quickly bulk up in Zone 3 to produce sizeable heads.

ZONE 3: SUNNY AND MOIST

Potatoes
Solanum tuberosum

	J F M A M J J A S O N D
CHIT	▬▬
PLANT	▬▬
HARVEST	▬▬▬▬

FROM PLANTING TO HARVEST
16–28 weeks

VARIETIES

'Lady Christl' – delicious creamy, early "new" potato

'Charlotte' – brilliant salad "new" potato, also bulks up for storage

'Sarpo Mira' – maincrop variety, slug and late blight resistant

'Jazzy' – tasty yellow-skinned salad variety, great for containers

ORIGINS
Sunny and warm central and southern America, in open soils with ample moisture

ALSO GROWS IN
Zones 1, 4, and 5

HARDINESS
Not hardy – killed by frosts

LIFE CYCLE
Perennial, grown as an annual

YIELD
★★★★/☆☆☆☆

EASY TO GROW
8/10

Tubers bulk up well in warm, moist soils (left). Harvest as new potatoes (centre) or leave maincrop types to attain a large size (right).

Hugely important economically, potatoes are a staple throughout temperate regions. Lift young tubers for deliciously sweet, melting "new" potatoes, or allow to bulk up for roasting, mashing, frying, or baking.

Growing essentials

Chitting Buy certified "seed" potatoes late winter or early spring, and place them in trays in a frost-free, well-lit spot to "chit" – develop shoots.

Planting Plant tubers into well-prepared beds, spacing individual tubers 30–40cm (12–16in) apart, or single tubers per 30cm (12in) diameter pot.

Harvesting Mound over soil- and pot-grown plants periodically as shoots emerge to encourage tubers. Water potted plants regularly. Harvest once tubers have reached desired size.

Secrets of success

Early potato varieties benefit from chitting, to hasten maturity and harvest.

Protect green shoots from frosts using old blankets or horticultural fleece held aloft with wire frames.

If late blight is problematic, grow blight-resistant varieties and harvest promptly.

When harvesting, slide a garden fork vertically alongside plants, then gently lever up tubers.

Prepare maincrop varieties for storage by allowing tubers to dry for a few hours after harvest.

Grow "second cropping" potatoes for winter by planting tubers in summer.

Lettuces
Lactuca sativa

	J	F	M	A	M	J	J	A	S	O	N	D
SOW			▬	▬	▬	▬	▬	▬				
TRANSPLANT				▬	▬	▬	▬	▬	▬			
HARVEST					▬	▬	▬	▬	▬	▬	▬	

FROM SOWING TO HARVEST
5–14 weeks

VARIETIES
'Catalogna' – vigorous oak-leaf, loose green head, stands well

'Little Gem' – benchmark, cos-style, easy, excellent quality

'Lobjoit's Green Cos' – delicious, large, sweet, pale green cos

'Gustav's Salad' – beautifully soft butterhead lettuces, exceptional quality

ORIGINS
Initially cultivated in north Africa, now widely grown in temperate and tropical regions

ALSO GROWS IN
Zones 1, 4, and 5

HARDINESS
Half to fully hardy

LIFE CYCLE
Annual

YIELD
★★★★/☆☆☆☆

EASY TO GROW
8/10

Lettuces offer a huge array of colours and textures. Quick to crop, leaves can be ready in just 5 weeks. Mild, succulent, and delicious with textures ranging from tender and buttery to full of fresh crunch.

Growing essentials

Sowing Sow seeds 1cm (½in) deep in modules under cover for early crops of heading lettuces. Sow loose-leaf and cut-and-come-again types in drills, in growing bags, troughs, or pots.

Planting Transplant heading varieties at 20–30cm (8–12in) spacings once large enough, watering well. Move container crops outside as weather warms. Lay organic slug pellets.

Harvesting Snip cut-and-come-again crops when leaves are usable, then water and liquid feed for subsequent harvests. Remove outer leaves of loose-leaf types, or cut headed forms, as required.

Secrets of success

Sow little and often to ensure crop succession, especially for heading types.

Plug plants are handy for non-sowers. Stagger by planting a few 2 weeks later.

Use cloches in a sheltered spot for an early crop – for seedlings started off under cover.

Summer sowings run to seed if allowed to dry out – be vigilant with watering.

Sow hardy heading types in August for winter harvests, or cut-and-come-again in September.

If bolting is a problem, choose loose-leaf types and sow in a shadier spot.

Lettuces will readily bulk up in sunshine provided ample moisture is available.

ZONE 3: SUNNY AND MOIST

Dahlias
Dahlia species

	J	F	M	A	M	J	J	A	S	O	N	D
SOW				▬	▬							
TRANSPLANT						▬						
HARVEST										▬		

FROM SOWING TO HARVEST
36–38 weeks

VARIETIES

'Black Jack' – burgundy cactus dahlia. Up to 1.5m (5ft) high.

'Wootton Impact' – yellow and orange cactus dahlia, tall and vigorous

'Buga Munchen' – pink cactus dahlia, vigorous, stocky

'Moonfire' – compact form, yellow blooms, ideal for pots

ORIGINS
The mountains of southern America and Mexico

ALSO GROWS IN
Zone 1

HARDINESS
Not hardy – killed by frosts

LIFE CYCLE
Perennial, grown as an annual

YIELD
★★★/☆☆☆☆

EASY TO GROW
8/10

As well as being highly ornamental, dahlias will yield tasty tubers for late-season harvests.

Their outstanding array of colours and late-season interest has made dahlia flowers an eternally popular ornamental, but they were first cultivated for their succulent edible tubers, now enjoying a revival.

Growing essentials

Starting Usually started off as tubers, in spring. Pot into individual 25cm (10in) diameter containers. Gently water and feed, in warmth, to encourage growth.

Planting Harden off and plant out when risk of frost has passed. Alternatively raise from seed, under cover in spring, or directly sow mid-April to mid-May.

Harvesting Pinch out to encourage bushiness. Stake taller varieties, and enjoy the flowers on plants or in a vase. Lift tubers once light frosts have blackened foliage late autumn.

Secrets of success

Seed-raised dahlias offer a surprisingly productive, cost-effective crop.

Mulch with rotted organic matter and irrigate during dry spells for best yields.

Tuber eating quality often improves in storage, from mild and juicy to sweet and fluffy.

To eat tubers peel off the skins, which can be tough or bitter, then consume raw or cooked.

Tubers are high in inulin, so introduce them slowly into your diet - to prevent bloating.

Store somewhere dry and frost-free. Packing in dry sand or compost deters withering.

ZONE 3: SUNNY AND MOIST

Pears
Pyrus communis

	J F M A M J J A S O N D
PLANT	▬▬▬ ▬▬
HARVEST	▬▬

FROM PLANTING TO GOOD HARVEST
1–4 years

VARIETIES

'Beth' – productive, early season, melting flesh, great flavour

'Conference' – mid to late season, sets fruit without a pollinator

'Williams' Bon Chrétien' – reliable, mid-season, tender, sweet flesh

'Doyenné du Comice' – late-season variety for storage, delicious and juicy

ORIGINS
Native to Europe, domesticated in Australia and North America

ALSO GROWS IN
Zone 1 and Zone 5

HARDINESS
Fully hardy, but flowers vulnerable to frost

LIFE CYCLE
Perennial

YIELD
★★★/☆☆☆☆

EASY TO GROW
7/10

Sinking your teeth into a perfectly ripe pear will persuade you to earmark a spot for these trees. Once mature, sizeable autumn harvests roll your way, with delicious late-season varieties extending that treat well into winter.

Growing essentials

Planting Plant in a sunny position, in well-drained soil dug over and improved with garden compost or well-rotted manure. Can be free-standing, or trained against a wall or fence.

Pruning Prune in winter, trained trees also appreciate summer pruning. Build up a framework of stems initially. At maturity remove a proportion of old wood and thin out fruiting spurs.

Harvesting Pear skin subtly yellows at maturity, and fruits part readily from spurs. Eat early varieties straight away, and let late-season fruits mellow in storage.

Secrets of success

Pear trees readily form spurs so lend themselves to cordon or espalier training.

Check pollination groups before purchase. Some pears are self-fertile but many require a pollination partner.

Pears flower early – avoid planting in frost pockets, or choose late-flowering types.

Pears are hungry feeders – apply a top dressing of blood, fish, and bone annually.

Pear 'Conference' thrives in warmth and a moist soil with good drainage.

Store late-maturing pears in a cool shed, garage, or fridge. Move batches into warmth to ripen.

Winter prune mature trees by reducing congested upright shoots.

ZONE 3: SUNNY AND MOIST

Florence fennel
Foeniculum vulgare var. *azoricum*

	J F M A M J J A S O N D
SOW	▬
HARVEST	▬▬▬▬

FROM SOWING TO HARVEST
12–20 weeks

VARIETIES
'Dragon' F1 – vigorous hybrid, heat and bolt resistant

'Rondo' F1 – strong-growing hybrid, quickly forms "bulbs"

'Zefa Fino' – Swiss-bred. Very cold and bolt resistant

'Mantovano' – Specifically bred for sowing earlier in spring, bolt resistant

ORIGINS
Fennel is native to the Mediterranean basin – with Florence fennel domesticated from this wild type

ALSO GROWS IN
Zone 1

HARDINESS
Not fully hardy – killed by harsh frosts

LIFE CYCLE
Biennial, grown as an annual

YIELD
★★★/☆☆☆☆

EASY TO GROW
7/10

Florence fennel, also known as bulb fennel although it's not a true bulb, is a delicious and highly ornamental plant. The swollen leaf bases have a characteristic aniseed flavour, as does the foliage. In warm locations with ample moisture, a row will deliver yields throughout autumn and into winter.

Growing essentials

Sowing Sow directly into soil for optimum performance. Make a drill 3cm (1¼in) deep, water the base then sow seeds thinly along its length. Cover over with soil and firm down gently.

Maintaining Water if conditions are dry. Once emerged, thin seedlings to 20cm (8in) apart – you can eat these. Keep the area weed-free and ensure soil remains moist throughout growth.

Harvesting Eat bulbs at any stage, though they are bulkiest come autumn. Slice the bulbs off at soil level to harvest.

Secrets of success

Bolt-resistant varieties are best for early sowings as cold spells trigger flowering.

Florence fennel dislikes transplanting, so starting off seeds in pots is possible, but can encourage bolting.

A long, steady supply of moisture is key for large bulbs.

The foliage is edible too – harvest occasional stems as bulbs mature.

If plants bolt the leaves are still edible, as are the flowers (which are also loved by pollinators).

Cover rows with cloches as winter arrives to maintain quality as long as possible.

F1 hybrid 'Rondo' bulks up very well when sown in Zone 3.

Ten other star performers

Melons
Two species in particular, *Cucumis melo* (which includes cantaloupes and honeydews) and *Citrullus lanatus* (watermelon) yield well outdoors. Early-to-mature and cold-weather varieties, like 'Alvaro' and 'Minnesota Midget', are particularly reliable in temperate climates.

Indian shot
This tender perennial hailing from the Americas is worth growing for its striking good looks, alone. The great news is that you can also consume the young starchy rhizomes of *Canna indica* raw or baked. Foliage can be used like banana leaves.

Sweetcorn
Quick-growing in warmth, sweetcorn (*Zea mays*) yields best with a strong root system – encourage this with early plantings or direct sowings, protected by cloches. For maximum cobs, ensure ample water when silks and tassels (flowers) appear.

Hyacinth bean
Vigorous tropical climber *Lablab purpureus* is tender but grows quickly in warm conditions. It enjoys moisture, yet dislikes waterlogging. Produces fragrant lilac flowers late summer, followed by pods best cooked and eaten young before seeds appear.

Mooli
Daikon, botanically *Raphanus sativus* 'Longipinnatus', a half-hardy radish, yielding exceptionally large, long pure white roots. An Asian biennial, grown as an annual. Lift in autumn, defoliate, and eat fresh or store somewhere frost-free until required.

Yacon

Earth apple or *Smallanthus sonchifolius*, a highly ornamental tuberous perennial from the Andean mountains, bearing large, arrow-shaped leaves covered in fine hairs. In late autumn, before the first frosts, lift and eat the large crunchy tubers.

Cucumber

A summer greenhouse stalwart, cucumbers also yield outside. These tender annual trailing or climbing plants (*Cucumis sativus*) enjoy shelter, and thrive in moist yet free-draining soil. Ridged types like 'Marketmore' are especially productive.

Spring onions

Flavoursome hardy plants (*Allium cepa*) offering pullings well before bulb onions and shallots. Best sown direct, some yield fresh bulbs, like 'Early Paris White', and 'Lilia', while others, 'Ishikura' and 'Kyoto Market', produce edible stems.

Dill

Beautifully aromatic Mediterranean herb *Anethum graveolens*. A hardy annual quickly growing from sowings into upright plants with feathery foliage and hints of warm aniseed. Stake once tall. Produces starbursts of yellow flowers, then edible seeds.

Shiso

Perilla frutescens, borderline hardy perennial, grown as annual. Wonderfully warm and aromatic basil and mint flavour and attractive foliage. Leaves often used as a vessel for Asian food. *P. f.* var. *crispa* has striking crinkled purple leaves.

Pictured from left to right, cucumber, Indian shot, shiso, and hyacinth bean all thrive in sunny sites with ample moisture.

Project
Waving goodbye to weedy plots

Weeds can steal the joy of growing your own food – don't let them! There are ways to keep on top, and they don't need to be banished, just managed so that crops aren't compromised. The warm, moist soils of Zone 3 offer perfect conditions for abundant growth – let's make it predominantly edible.

We divide weeds into two groups: annual and perennial. Each has its own strengths – taking action against these is key to effective control.

Annual weeds

Annuals self-seed quickly and freely. If allowed to, populations explode. Examples include chickweed (*Stellaria media*), bittercress (*Cardamine hirsuta*), ragwort (*Senecio jacobaea*), and fat hen (*Chenopodium album*). The potential for growth is almost unfathomable – one chickweed plant can produce 2,500 viable seeds in 10 weeks, so by summer's end, 15 billion seeds could be set! Removing plants via hoeing or hand weeding before they seed is crucial. Hoeing is most effective on sunny, breezy days – shallow-hoe with a sharp tool to slice weeds off at soil level. Don't bury the weeds, or smear them with a blunt hoe – they will grow back.

Perennial weeds

These weeds have persistent, aggressive roots – examples include bindweed (*Convolvulus arvensis*), ground elder (*Aegopodium podagraria*), perennial nettle (*Urtica dioica*), bramble (*Rubus*), and horsetail (*Equisetum arvense*). The roots are more extensive on open, lighter soils – horsetail roots can reach 2m (6½ft) deep. Digging out roots can control shallow-rooted weeds like bramble, nettle, and ground elder, but this may be temporary if you border an inaccessible area of weeds that can simply reinvade. Sink paving slabs or weather board horizontally to a depth of 50cm (20in) along these boundaries.

ZONE 3: **SUNNY AND MOIST**

You will need

Sheets of cardboard
Bricks or similar weighty items

Steps

1 Clear the bed of existing weed growth.

2 Lay the cardboard over the soil.

3 If working around individual or permanent plants, cut holes around the base of crops so the cardboard sits snugly around them. Make slits for drills of crops.

4 Fix the cardboard in place using weighty bricks.

5 You can opt to hide the cardboard with a gravel or bark mulch. Weeds may eventually poke through holes or germinate into the mulch. Remove them before they take hold.

Extreme conditions such as these have connotations of stark growth and poor yields. Luckily there are plenty of edibles that think otherwise! Temperate climates have allowed the evolution of supremely hardy crops. Think of winter plots rich with kale, cabbage, and Brussels sprouts. Leeks and chicories, too, can hunker down, ready to be prised from frozen soils. Hardy root crops form the backbone of exposed plots – carrots, parsnips, swede, beets, and turnips can all ride out sub-zero temperatures when buried beneath the earth. Fruits are happy to suffer the cold, too, with damsons, quinces, raspberries, and cranberries just some of the sweet treats on offer. A zone of scarcity, this is not.

Zone 4
Open and cold

ZONE 4: OPEN AND COLD

Onions and shallots
Allium cepa

	J	F	M	A	M	J	J	A	S	O	N	D
SOW			▬	▬	▬							
TRANSPLANT				▬	▬						▬	▬
HARVEST							▬	▬	▬			

FROM SOWING TO HARVEST
18–26 weeks

VARIETIES
'Zebrune' – pink, sweet, heritage banana-shallot

'Armstrong' F1 – classic round Rijnsburger onion, keeps excellently

'Red Tide' F1 – vigorous, glossy, round, red-skinned, excellent storage

'Snowball' – white, mild-tasting bulbs, stores well, often sold as sets

ORIGINS
Asia, but adaptable to a wide range of conditions

ALSO GROWS IN
Zones 1, 3, and 5

HARDINESS
Fully hardy

LIFE CYCLE
Biennial, grown as an annual

YIELD
★★★★/☆☆☆☆☆

EASY TO GROW
7/10

A key ingredient in innumerable dishes, home-grown onions store excellently for year-round supplies. Shallots, often said to be more refined in quality, also store brilliantly. Cooked, their sweet, umami flavours taste divine.

Growing essentials

Sowing Sow seeds 1cm (½in) deep, in modules under cover in spring. Grow until large enough to plant out. Buy sets (tiny bulbs) if sowing isn't possible.

Planting Plant sets or hardened off seedlings at 20–25cm (8–10in) spacings. Water and weed well whilst young. Watch that sets don't dislodge themselves.

Harvesting Eat bulbs when large enough. Allow to grow to maturity with hardened outer skins before storing somewhere airy and frost-free.

Secrets of success

Sow seeds individually for larger bulbs – or in clumps of 3–4 for smaller bulbs.

Acid soils may yield poorly – lime before planting to raise pH levels.

If allium leaf miner, onion fly, or leek moth are local, cover beds with insect-proof mesh.

Space onions according to your needs – closer for smaller bulbs, and vice versa.

To deter bolting choose heat-treated sets, avoid planting early, and water in dry spells.

Shallots (top) and bulb onions (bottom) offer cold-tolerant harvests with an excellent storage life.

Plait bulbs into ropes for space efficiency. Do this after 2 weeks of storage.

ZONE 4: OPEN AND COLD

Carrots and parsnips

Daucus carota and *Pastinaca sativa*

	J F M A M J J A S O N D
SOW	▬▬▬▬
HARVEST	▬▬▬ ▬▬▬▬▬▬

FROM SOWING TO HARVEST
14–28 weeks

VARIETIES

Carrot 'Amsterdam Forcing 3' – useful for forcing early baby roots

Carrot 'Extremo' F1 – very hardy, sow early summer for winter roots

Parsnip 'Viper' F1 – hardy, vigorous, large, uniform roots

Parsnip 'Pearl' F1 – hardy, uniform, slender white roots

ORIGINS

Eurasia, in deep, rich soils with steady rainfall

ALSO GROWS IN

Zone 1, Zone 3, Zone 5, and Zone 6 – in sun or shade, as long as soil isn't dry

HARDINESS

Hardy, some extremely hardy

LIFE CYCLE

Biennial, grown as annual

YIELD

★★★★/☆☆☆☆

EASY TO GROW

6/10

Winter soups, roasts, and casseroles wouldn't be complete without these two root crops. Rich in sugars, they bulk up well in the moisture of autumn to stand firm come winter.

Growing essentials

Sowing Sow directly into soil. Make drills 2cm (¾in) deep for carrots, 4cm (1½in) deep for parsnips, sow thinly along base, cover with soil and water in. For winter roots, sow carrots in May or June.

Planting Keep beds weed-free. Thin resulting seedlings to 4cm (1½in) apart for winter carrots, 20cm (8in) apart for mature parsnips. Water during dry spells.

Harvesting Pull baby carrots in June for early crop, otherwise leave to bulk up into late autumn. Pull as and when required – a garden fork may be needed to ease parsnips out.

Secrets of success

Parsnip seeds are short-lived – buy fresh each year.

Germination can take 2–3 weeks. Sow parsnips May onwards, in warm soil.

Carrot fly is easily prevented with insect-proof mesh.

Insulate roots in winter by removing leaves, then mulching with straw pegged down under netting.

Frosts boost parsnips' sugar levels, as starches convert into anti-freezing sugars.

Cover arms when harvesting parsnips – the foliage can irritate sensitive skin.

Both carrots (left) and parsnips (right) are versatile edibles with excellent hardiness.

Broad beans
Vicia faba

FROM SOWING TO HARVEST
20–30 weeks

VARIETIES
'Aquadulce Claudia' – very hardy, autumn sowing, large plants and pods

'De Monica' – matures quickly, spring sowing, good-sized pods

'The Sutton' – dwarf variety, spring sowing, short pods, can grow in pots

ORIGINS
Exact origins unknown, possibly Middle East. Hardier than most beans so very useful as early summer crop

ALSO GROWS IN
Zone 1, Zone 3, and Zone 5

HARDINESS
Half to fully hardy

LIFE CYCLE
Annual

YIELD
★★★/☆☆☆☆☆

EASY TO GROW
8/10

High quality broad beans are difficult to buy for the kitchen; you'll harvest fresher, more tender beans at home. Any gluts resulting from the short season are forgivable as broad beans freeze well.

Growing essentials

Sowing Sow seeds in large modules or root trainers, under cover. Transplant outside once large enough. Alternatively, sow directly into soil, 2–3cm (¾–1¼in) deep, in grids. Space plants 20 × 20cm (8 × 8in) apart.

Planting Keep beds weed-free until crops establish. Water in dry spells, especially when flowering. Encircle with canes and string as plants grow.

Harvesting Pick when beans are fully formed but not too old. Remove beans from pods and eat as soon as possible.

Secrets of success

Cooler conditions give best germination, so you won't need a heated propagator.

Avoid sowing late in the season when temperatures rise – these beans yield poorly compared to French and runner beans.

Autumn sowings overwinter best if plants are 3–5cm (1¼–2in) tall – so don't sow in haste as loftier plants often snap.

Heavy wet soils rot autumn sowings – avoid by sowing in pots under cold frames.

Snap out plant tips in late spring to deter black bean aphids – tips can be steamed and eaten.

Broad bean plants can be supremely hardy, developing clusters of attractive spring flowers (top), followed by plump pods (bottom).

Roots develop nitrogen-fixing nodules – leave them in the soil when removing spent plants.

Plums, damsons, bullaces, and sloes

Prunus species

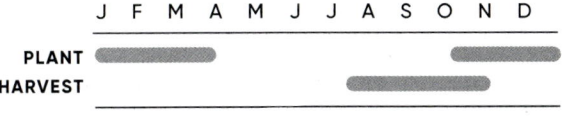

PLANT
HARVEST

J F M A M J J A S O N D

FROM PLANTING TO HARVEST
1–4 years

Plums and their relatives are one of the more reliable fruits in exposed conditions, frequently setting heavy crops.

VARIETIES

Plum 'Victoria' – excellent, self-fertile, reliable

Damson 'Merryweather Damson' – good-sized purple fruits, self-fertile, reliable, hardy

Bullace 'Shepherd's Bullace' – yellow culinary fruits, hardy, self-fertile

Sloe 'Blackthorn' – very hardy, suckering, self-fertile, small, sour, purple fruits

ORIGINS
Ancestors native to Europe and Asia

ALSO GROWS IN
Zone 1, Zone 2, Zone 3, Zone 5, Zone 6, and Zone 7

HARDINESS
Fully hardy, flowers vulnerable to frost

LIFE CYCLE
Perennial

YIELD
★★★★/☆☆☆☆

EASY TO GROW
8/10

Fruit size varies within this group, but their desire to provide you with harvests is relentless. Easy to grow and robust, in good years you can be inundated with fruits. They can be preserved in all manner of delicious ways, so welcome gluts with open arms.

Growing essentials

Planting Plant in soil improved with garden compost or well-rotted manure. If fan training plums or gages erect wires or trellis.

Pruning Prune plums in spring and summer as winter pruning invites disease. At maturity remove a proportion of old wood to keep plants productive. Sloes, bullaces, and damsons are happy unpruned.

Harvesting Harvest plums, damsons (and gages) when fully mature for the best juice and flavour. Leave bullaces and sloes on the tree for as long as possible.

Secrets of success

Sloes, bullaces, and damsons crop most reliably in cold regions.

A warm spot will boost sugar levels of plums and gages.

Trees flower early in spring – to guarantee heavy yields avoid planting in frost pockets.

If fan training plums or gages prune out a proportion of old stems and tie in new, each summer immediately after harvest.

Prop up branches if fruit set is heavy.

Thwart plum moth damage by hanging pheromone traps in late April and May.

Beetroot

Beta vulgaris

	J	F	M	A	M	J	J	A	S	O	N	D
SOW			▬	▬	▬	▬						
TRANSPLANT				▬	▬	▬	▬					
HARVEST						▬	▬	▬	▬	▬		

FROM SOWING TO HARVEST
10–16 weeks

VARIETIES

'Pablo' F1 – vigorous, uniform, purple-rooted form, excellent yields

'Boldor' F1 – supreme golden beets. Orange skin, yellow flesh, very tasty

'Barbabietola di Chioggia' – exceptionally eye-catching purple- and white-ringed roots

'Avalanche' F1 – vigorous, unusual, supremely sweet white-rooted beet

ORIGINS
Domesticated in Egypt for its leaves, and now grown for roots. Enjoys a well-drained soil and a temperate climate

ALSO GROWS IN
Zones 1, 3, and 5

HARDINESS
Fully hardy

LIFE CYCLE
Biennial grown as an annual

YIELD
★★★★/☆☆☆☆☆

EASY TO GROW
8/10

Newly appreciated for its health benefits, beetroot can be pickled, roasted, and juiced. The crop is one of the quicker roots to mature, with improved breeding giving ever-increasing vigour and colour ranges.

Growing essentials

Sowing Sow seeds 1cm (½in) deep, in modules, under cover in spring, or directly into freshly prepared and well-raked ground, outside, as spring temperatures rise.

Planting Thin direct-sown seedlings to 12–20cm (4¾–8in) apart, depending on your required root size. Harden off and transplant modules at similar spacings once large enough to handle. Water in well.

Harvesting Weed rows well, water during dry spells, then harvest when required. Baby beets are deliciously tender. Larger roots are ideal for winter storage.

Secrets of success

For single large roots, buy "monogerm" varieties or thin seedlings to one per station.

Multi-sowing or growing in clumps gives smaller roots.

Beetroot leaves are edible – consume thinnings, harvest occasional larger leaves.

Waterlogging isn't tolerated – grow in raised beds on poorly drained plots.

Enjoying slow, steady moisture the large, robust roots of beetroot can be stored in sand for out-of-season use.

Water during dry spells to improve quality, yield, and reduce bolting risk.

Store mature roots in boxes of just damp sand, in a frost-free shed or garage.

Winter kales and cabbages
Brassica oleracea

	J F M A M J J A S O N D
SOW	▬▬
TRANSPLANT	▬▬
HARVEST	▬▬▬▬▬ ▬▬▬▬▬

FROM SOWING TO HARVEST
20–24 weeks

VARIETIES
Cabbage 'Tundra' F1 – excellent savoy, hardy, bulks up quickly, stands well

Cabbage 'Deadon' F1 – robust, hardy January King type, red-tinged leaves

Kale 'Cavolo Nero' – Italian, heirloom, ribbed, rich green leaves

Kale 'Daubenton's' – robust, perennial, gently scalloped mid-green leaves

ORIGINS
Domesticated in Europe, many varieties show extreme hardiness

ALSO GROWS IN
Zones 1, 3, and 5

HARDINESS
Hardy

LIFE CYCLE
Biennial or perennial

YIELD
★★★★/☆☆☆☆

EASY TO GROW
6/10

Less robust varieties are available for cropping in the warmer seasons, but hardy kales and cabbages, boosted by autumn rainfall, offer good harvests during the leaner months. Started off in spring, they quietly bulk up, filling the winter void.

Growing essentials

Sowing Sow winter-heading crops directly into soil, or in modules one seed per cell, for later transplanting. Spacing at least 60cm (24in) apart.

Maintaining Keep well-hoed until established, and well-watered during dry spells. Protect from cabbage white butterflies under a cage of 5 × 5mm (¼ × ¼in) netting.

Hardy kale 'Cavolo Nero' stands well all winter to give sustained harvests.

Harvesting Harvest as soon as heads or leaves are large enough. A small flush of loose-leaf "greens" arises from cabbage stumps if left in the ground once harvested.

Secrets of success

If club root is troublesome, choose varieties with resistance – cabbage 'Kilaton' or kale 'Tall Green Curled'.

Add a high-nitrogen fertilizer like chicken pellets to the bed, just prior to planting.

Deter cabbage fly by placing brassica collars around young transplants.

Adjust butterfly netting as plants grow, to hold it away from leaves at all times.

Pigeons target crops in winter – ensure netting remains in place to thwart them.

Winter greens and spring cabbage are also very hardy – sow in autumn and early spring.

Blueberries
Vaccinium species

	J F M A M J J A S O N D
PLANT	▬▬ ▬▬
HARVEST	▬▬▬▬

FROM PLANTING TO HARVEST
1–3 years

VARIETIES
'Goldtraube' – large, tasty berries on vigorous bushes

'Pink Berry' – sweet pink-coloured fruits, robust healthy bushes

'Sunshine Blue' – compact plants, abundant yields, ideal for containers

'Jersey' – fruits later than most extending the fresh season, small berries

ORIGINS
Diverse, predominantly from temperate regions of the Americas

ALSO GROWS IN
Zones 1, 3, 5, and 6

HARDINESS
Hardy, some extremely

LIFE CYCLE
Perennial

YIELD
★★★/☆☆☆☆

EASY TO GROW
7/10

Renowned for health-boosting antioxidants, blueberries are a staple on any fruit grower's plot. Versatile enough to be container grown, their love of moisture and acidic growing conditions is well known. With careful selection, delicious, fresh blueberries can be yours from June until October.

Growing essentials

Planting Plant during their dormant period into well-drained yet moisture-retentive soil enriched with acidic organic matter. Water well and add a thick layer of mulch.

Pruning Prune bushes just before bud burst in spring. Simply shape young plants. Remove weak or congested wood on mature plants, opening canopies.

Harvesting Harvest berries once fully coloured up, gently rolling them off with your finger and thumb. Large batches are excellent for home freezing.

Secrets of success

Blueberries crop on their own but yield improves when planted with other varieties.

A planting pit lined with a pierced pond liner ensures soils remains moist.

Pot growing in ericaceous compost is ideal for chalky gardens.

Bird protection is essential when berries ripen – erect a wire mesh cage over plants.

Lay seep hoses under an organic mulch to irrigate larger beds.

Healthy sections of last year's growth make excellent hardwood cuttings when plants are pruned in winter.

Excellent hardiness make blueberries 'Jersey' (top) and 'Pink Berry' (bottom) ideal for cooler plots.

ZONE 4: OPEN AND COLD

Leeks
Allium porrum

	J F M A M J J A S O N D
SOW	▬▬▬
TRANSPLANT	▬▬▬
HARVEST	▬▬▬▬▬▬ ▬▬▬▬▬

FROM SOWING TO HARVEST
22–30 weeks

VARIETIES

'Warwick' F1 – early, vigorous, long white shanks, holds well into winter

'Porbella' – mid-season, very hardy variety, thick shanks, rust resistant

'Lancaster' F1 – good, late-season leek, very hardy, stands well, bolt resistant

'Oarsman' F1 – British bred, bulks up vigorously, and stands well through winter

ORIGINS
Cultivated in Asia and North Africa, in deep, fertile soils with good drainage

ALSO GROWS IN
Zones 3 and 5

HARDINESS
Hardy, some very

LIFE CYCLE
Biennial, grown as annual

YIELD
★★★★/☆☆☆☆

EASY TO GROW
7/10

Ubiquitous with winter veg plots, where it politely stands in all weathers, until needed in the kitchen. The beautifully mild stems can be sauteed in butter as a side dish, blended with potato as a soup, or mixed with bechamel sauce for a cool-season treat.

Growing essentials

Sowing Sow in seedbeds outdoors, or large pots under cover. Lift and separate out once stems are 2–4mm (1/16–1/8in) thick. Leeks can be multi-sown, 3–4 seeds per large module.

Planting Transplant individual seedlings 20cm (8in) apart – closer spacings will give baby leeks. Drop into holes 12–15cm (5–6in) deep. Plant clumps of multi-sown leeks 25cm (10in) apart, each way. Water well and keep weed-free.

Harvesting Water in dry spells, lift individual leeks as and when. Trim roots and uppermost green leaves before eating.

Secrets of success

Avoid leek rust in mild, wet locations with resistant varieties.

Late-season leeks with deep green foliage like 'Northern Lights' are incredibly hardy.

High-nitrogen feeds like poultry pellets, are beneficial. Dig in just prior to planting.

If leek moth is troublesome locally, prevent attacks with insect-proof mesh.

Leek 'Warwick' stands well throughout the winter months.

Keep beds weed-free – leeks are poor at competing due to their upright growth.

Uproot in batches and "plant" in a frost-free tub when frost is forecast, as lifting from frozen soil is difficult.

Turnips, swede, and kohlrabi

Brassica species

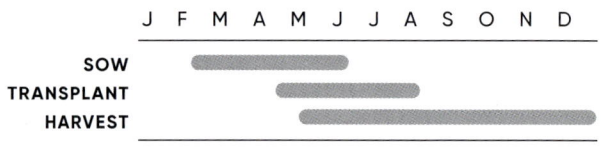

FROM SOWING TO HARVEST
6–18 weeks

VARIETIES
Turnip 'Sweet Marble' F1
– fast-growing, white, sweet

Kohlrabi 'Modrava' F1
– striking purple globes, quick-growing, succulent

Kohlrabi 'Quickstar' F1
– pale green, bulks up rapidly

Swede 'Tweed' F1
– British-bred, orange roots with purple shoulders

ORIGINS
Cool, temperate regions

ALSO GROWS IN
Zone 1, Zone 3, and Zone 5

HARDINESS
Hardy

LIFE CYCLE
Biennial, grown as an annual

YIELD
★★★/☆☆☆☆

EASY TO GROW
7/10

You can't beat a robust root crop in the winter months. Expect bulky harvests midsummer until late autumn, plus stored produce winter through spring from this trio. While baby turnips are quick to bulk up, others take their time to provide you with sizeable yields.

Growing essentials

Sowing Sow seeds 1cm (½in) deep, in modules, under cover in spring, or directly into freshly prepared ground, outside, as spring temperatures rise.

Thin direct-sown seedlings 12–30cm (4¾–12in), depending on the root size required. Harden off and transplant modules at similar spacings once large enough to handle, watering well.

Harvesting Weed well and water during dry spells. Harvest as and when required. Baby turnips and kohlrabi are deliciously tender; larger swedes are ideal for storage.

Secrets of success

For quick baby turnips sow under cover in March, planting under cloches in April.

Turnip 'Sweet Marble' (top) and kohlrabi 'Modrava' (bottom) are vigorous, quick cropping hybrids that easily shrug off chillier weather.

Swedes are best sown late May or early June, then left to bulk up for winter.

All three are brassicas – rotate into new beds if clubroot or cabbage fly are troublesome.

You can eat turnip foliage, taking a few leaves from developing plants.

Cover plants, especially kohlrabi, with 5mm (¼in) mesh to protect from cabbage white caterpillars.

Lift swede in autumn, remove leaves, and store in boxes of just damp sand for winter.

Apples
Malus domestica

	J F M A M J J A S O N D
TRANSPLANT	▬▬▬ ▬▬
HARVEST	▬▬▬▬

FROM PLANTING TO HARVEST
1–3 years

VARIETIES
'Cox Self Fertile' – robust, delicious, orange-flushed fruits. Only one tree needed

'Kidd's Orange Red' – vigorous, dessert apple. Exceptional red-flushed fruits

'James Grieve' – hardy dual-purpose apple. Tasty green fruits with red flush

'Bramley's Seedling' – Reliable, vigorous cooking apple. Large, mid-green fruits

ORIGINS
Mountainous Asia. Domesticated to grow in diverse conditions

ALSO GROWS IN
Zone 1, Zone 2, Zone 3, Zone 4, and Zone 5

HARDINESS
Hardy

LIFE CYCLE
Perennial

YIELD
★★★/☆☆☆☆

EASY TO GROW
8/10

Nothing matches the crunch of an apple, and home gardeners have access to exciting varieties. Welcome a robust russet, a fluffy cooker, or a juice-packed dessert. Dwarf rootstocks and self-fertile selections ensure compact plots aren't excluded.

Growing essentials

Planting Plant dormant trees in weed-free soil enriched with well-rotted organic matter. Mulch and stake until well established.

Pruning Shape young trees in winter. Remove congestion and crossing limbs in mature trees. Summer prune excess leafy growth on compact shapes.

Harvesting Harvest individual fruits as they ripen, picking over trees several times. Only pick fruits that part readily from the tree. Store in a cool, frost-free spot.

Secrets of success

Ask about flowering groups when purchasing, as this governs fruit set.

Some, like 'Discovery', are best eaten fresh. Others, like 'Blenheim Orange' improve with storage.

Buy trees with good framework branches. Bare-root plants are a cheaper winter option.

Rootstocks determine tree size. M25 is extremely dwarfing, MM106 gives a large tree.

Correct planting depth is essential for health – check hole depth using a baton.

Compact shapes like cordons, stepovers, and espaliers are ideal for wall training and smaller plots.

Apples (here, 'Kidd's Orange Red') hail from temperate climates and so they perform well in cooler conditions.

Ten other star performers

Mustards
A vast range of hardy annuals classified as *Brassica juncea*, mustards offer quick-to-mature harvests that are delicious. Thriving in cool, moist spring and autumn conditions, look for selections such as 'Green in Snow', 'Red Frills', 'Red Giant', and 'Wasabi'.

Chinese flowering quince
A suckering deciduous shrub (*Chaenomeles cathayensis*) to 3m (10ft). White flowers flushed pink appear in early spring. Native to high Himalayas, so very cold-hardy. Large, apple-shaped fruits, borne in late summer and early autumn. Best in sun though takes some shade.

Sea buckthorn
Incredibly hardy deciduous shrub, *Hippophae rhamnoides*. Small orange berries are best cooked. Produced in sizeable clusters on young stems of female plants. Grows to 4m (13ft) tall, found on Asian and European coasts, so tolerates dry soils.

Hardy peas
While wrinkle-seeded peas *Pisum sativum* are less cold tolerant, hardy, round-seeded, early-maturing varieties like 'Meteor', 'Douce Provence', and 'Feltham First' can be sown on lighter soils in autumn to overwinter, or early spring in pots, to transplant out onto heavier soils.

Brussels sprouts and cauliflowers
These two North Europe brassicas, both *Brassica oleracea*, show excellent cold-hardiness. Although space-hungry they are often the backbone of a winter plot. Provide nutrition, especially nitrogen.

ZONE 4: OPEN AND COLD 121

Korean celery

A perennial (*Dystaenia takesimana*) from temperate Asia. Supremely winter-hardy and vigorous. Finely cut foliage, tasting of lovage and leaf celery. Quickly grows from seed. Tolerates some shade. Huge white flowers, pollinator-friendly.

Arctic raspberry

A remarkably hardy fruit (*Rubus arcticus*) from Northern Europe. Makes excellent groundcover. Highly ornamental, edible, deep pink flowers, juicy, highly flavoured small raspberry fruits. Prefers moist, well-drained soil.

Cranberry and lingonberry

Two incredibly hardy evergreens (*Vaccinium oxycoccos* and *V. vitis-idaea*) from the Northern Hemisphere. Tart fruits on ground-hugging self-fertile plants. They originate from bogland, so moist, acidic soils are required.

Dog rose

A very hardy rose (*Rosa canina*) found widely in Europe. Makes an excellent informal hedge or specimen shrub. The single, pale pink flowers are loved by bees, and rosehips are an excellent source of vitamin C. Cook and strain these to make a syrup.

Chicory and endive

Excellent pickings through chillier months. Chicory, radicchio (*Cichorium intybus*), and endive (*C. endivia*) thrive in cool, moist conditions. Both found in Northern Europe. Can be blanched to reduce bitterness.

Pictured from left to right, sea buckthorn, chicory, Brussels sprouts, peas, and cauliflowers are all robust enough to hold their own on exposed plots.

Project: Planting an edible windbreak

Cold, open, and presumably inhospitable – well, thankfully not for all plants! Zone 4 can prove relentlessly brisk, but there are plenty of fruits that will enjoy it. The fruits, by nature, may be small but they give ample opportunity for foraging, preserving, and even home-brewing. Creating a windbreak makes life easier for other crops, too, as it filters gusts into a more steady breeze.

On open sites in a frost-prone location, your edibles must be hardy and robust. Luckily, many crops fit this description, and some may boost the diversity of your plot. Windbreak edibles are invaluable in that they offer relief from blustery conditions. They hold their own as a hedge, shield less robust crops, and you can harvest fruit from them, too. Why on earth would you not plant some?

The best time to establish a windbreak is between autumn and spring, when the soil is neither frozen nor waterlogged. Planting a mix of species will offer the most versatile and prolonged harvest (see list, opposite below). Often these are available bare-rooted in winter. Staking isn't practical, so begin with younger, smaller plants, which will be cheaper, and let them grow and knit together over the years. Many of these edibles have additional benefits. The flowers provide valuable food for foraging bees, the branches shelter for insect-feeding birds. Any fruit not taken yourself will be much appreciated by migratory fieldfares and redwings come winter.

ZONE 4: OPEN AND COLD

You will need

Fork or spade
Slow-release fertilizer
A selection of windbreak edibles
Temporary fencing (if rabbits are a problem)

Steps

1 Dig a planting trench wide and deep enough to accommodate your chosen plants' rootballs.

2 Mix some slow-release fertilizer into the base of the trench.

3 Work this fertilizer into the soil to distribute it.

4 Place your plants in a row, spacing them 30cm (12in) apart.

5 Plant your stock, working soil well between the roots. Firm in gently and water well. If rabbits are troublesome in your locality, erect temporary fencing until established.

Windbreak edibles:

Hawthorn (*Crataegus*)
Damson (*Prunus domestica* subsp. *insititia*)
Sloe (*Prunus spinosa*)
Dog rose (*Rosa canina*)
Rugosa rose (*Rosa rugosa*)
Elder (*Sambucus nigra*)
Serviceberry (*Amelanchier*)
Crab apple (*Malus sylvestris*)
Bird cherry (*Prunus padus*)
Cherry plum (*Prunus cerasifera*)
Gorse (*Ulex europaeus*)
Sea buckthorn (*Hippophae rhamnoides*)
Autumn olive (*Elaeagnus umbellata*)
Barberry (*Berberis*)

This goldilocks zone of not too sunny, not too shady, is perfect for edibles that enjoy life in the middle lane. Speedy leafy veg like rocket, coriander, and spinach are well known for quickly running to seed if they experience hot, dry spells – not so in the cool, balanced conditions of Zone 5. Plant with them an abundance of less commonplace leaves – Texsel greens, rapini, garlic cress, brighteyes, komatsuna, pak choi, and tatsoi – to stretch out the season of healthy, home-grown greens. Fruits, too, will enjoy these intermediate conditions. Raspberries and elderberries can be partnered with Chilean guavas, fuchsia berries, and juneberries. Cool, moist roots are also favoured by herbs such as mint and lovage, and warm spices like Szechuan pepper. This mid-zone is a popular place to be.

Zone 5
Part shade

Salad rocket

Eruca vesicaria subsp. *sativa*

	J	F	M	A	M	J	J	A	S	O	N	D
SOW			▬	▬	▬							
TRANSPLANT				▬	▬							
HARVEST					▬	▬	▬	▬	▬	▬	▬	

FROM SOWING TO HARVEST
6–10 weeks

VARIETIES

'Apollo' – large, lush, milder taste without bitterness

'Dragon's Fire' – red veins, finely cut green leaves

'Wasabi Rocket' – peppery, wasabi-like taste

'Astra' – very finely cut leaves, less bolting

ORIGINS
Mediterranean, loving warmth and some shade

ALSO GROWS IN
Zone 1, Zone 3, Zone 4, and Zone 6

HARDINESS
Hardy

LIFE CYCLE
Annual

YIELD
★★★★★/☆☆☆☆☆

EASY TO GROW
9/10

Many have tucked into a plateful of rocket, eager to try something different to milder salads. The peppery flavour doesn't disappoint, and the speed of growth sits well with gardeners. Soup, pesto, and hummus are all options – feel free to experiment with this versatile leaf.

Growing essentials

Sowing For an early crop, sow 2–3 seeds per cell, under cover in modules. Transplant outside 15cm (6in) apart under cloches early to mid-spring. Protect from slugs and snails.

As soil conditions warm in spring, sow directly into soil. Excavate a drill 5cm (2in) deep, and sow 1–2cm (½–¾in) apart. Thin (and eat) emerging seedlings to 12–15cm (4¾–6in) spacings.

Harvesting Harvest leaves when large enough – leave sufficient foliage so that plants can re-sprout for further pickings, keeping the growing point intact.

Secrets of success

Flowers are edible, but leaves from bolted plants are very peppery.

Hot, dry weather gives overly hot flavours – provide some shade mid-summer.

Grow year-round in mild gardens – sowing in autumn, for winter harvests.

Rocket grows well sown into trays on windowsills as a cut-and-come-again microleaf (see page 176).

For a low-maintenance option, leave plants to self-seed around the plot.

Look for wild rocket, *Diplotaxis tenuifolia,* a hardy perennial with a long harvest.

Batches of lush rocket leaves can be harvested from sowings as and when needed in the kitchen. Sowings made in the shade are less likely to run up to flower.

ZONE 5: PART SHADE

Annual spinach
Spinacia oleracea

	J	F	M	A	M	J	J	A	S	O	N	D
SOW				▬▬▬▬								
TRANSPLANT					▬▬▬▬							
HARVEST						▬▬▬▬▬▬						

FROM SOWING TO HARVEST
8–14 weeks

VARIETIES

'Apollo' F1 – quick-growing, ideal for early crops, excellent disease resistance

'Rubino' F1 – eye-catching, red stems and veins on green leaves

'Winter Giant' – supremely hardy, large-leaved, ideal for late-season sowings

'Medania' F1 – very good bolting resistance, perfect for summer cropping

ORIGINS
The Middle East – thriving as a cool season crop

ALSO GROWS IN
Zones 1, 3, 4, and 6

HARDINESS
Hardy

LIFE CYCLE
Annual

YIELD
★★★★⯪☆☆☆☆☆

EASY TO GROW
7/10

With rich green leaves, quick growth, and deliciously velvety texture, annual spinach is popular on plots worldwide. It grows in a variety of places and, given ample moisture, will provide generous, vitamin-packed harvests. Fresh or cooked, it's an essential green veg.

Growing essentials

Sowing Sow under cover in modules for an early crop. Sow 2–3 seeds per cell, and transplant outside 15cm (6in) apart, under cloches early to mid-spring. Protect from slugs and snails.

As soil conditions warm in spring, sow directly into beds. Excavate a drill 5cm (2in) deep, and sow seeds 1–2cm (½–¾in) apart. Thin (and eat) emerging seedlings to 12–15cm (4¾–6in) apart.

Harvesting Harvest leaves when large enough – leave enough foliage so that plants can re-sprout for further pickings. Alternatively, harvest whole plants.

Secrets of success

Transplant early crops in a sunny, sheltered spot to make use of any spring warmth.

Sow very hardy types late summer, for pickings throughout winter.

Gluts freeze well – wilt down in seasoned butter or oil, then compress.

Shady spots offer protection from summer heat, which causes bolting.

Sow little and often for harvests through the growing season.

The crop is happy in containers or growing bags – water plants well.

Annual spinach can be sown little and often for a succession of fresh leaves – shade offers relief from hot summer sun.

ZONE 5: PART SHADE

Raspberries
Rubus idaeus

	J F M A M J J A S O N D
PLANT	▬▬ ▬▬
HARVEST	▬▬▬▬

FROM PLANTING TO HARVEST
6 months–2 years

These incredibly productive plants are very versatile, thriving in a variety of conditions – dwarf forms are even available for growing in pots. The fruits themselves are unrivalled in flavour and gluts are easy to accommodate in the kitchen by any keen preserver.

VARIETIES
'Tulameen' – vigorous, productive, excellent summer variety

'Ruby Beauty' – compact summer variety, to 50cm (20in) tall

'Fallgold' – striking, yellow-fruited autumn raspberry, vigorous, thorny canes

'Polka' – productive autumn raspberry, yielding until the first frosts

ORIGINS
Across Europe, enjoying the cool, moist conditions of autumn and spring

ALSO GROWS IN
Zone 1, Zone 3, Zone 4, and Zone 6 – if not waterlogged

HARDINESS
Hardy

LIFE CYCLE
Perennial

YIELD
★★★★/☆☆☆☆☆

EASY TO GROW
8/10

Growing essentials

Planting Plant canes when dormant into weed-free soil enriched with well-rotted organic matter. Summer types are best supported by a wall, fence, or a system of wires and posts.

Pruning Mulch plants annually in spring. Prune autumn varieties to soil level each winter. Remove fruited canes of summer types in autumn, and tie new canes in to take their place.

Harvesting Harvest fruit once it is fully coloured, gently rolling it off the plant between finger and thumb. Pick over plants every day or so as fruits ripen individually.

Secrets of success

Summer raspberries, dubbed "floricane", crop on one-year-old canes which then die.

Autumn raspberries or "primocane" types, crop on the current year's canes.

Little white grubs in ripe fruits are indicative of raspberry beetles. Reduce with pheromone traps.

Raspberries prefer acidic soils, otherwise magnesium deficiency occurs – rectify this with a foliar feed of Epsom Salts.

Thin out excess new and wayward canes, once well emerged in early summer.

Eat very young shoots while tender, and steep leaves as a herbal tea.

Raspberries 'Fallgold' (top) and 'Tulameen' (bottom) will yield very well in the shadier area of your garden.

ZONE 5: PART SHADE

Elder
Sambucus nigra

	J	F	M	A	M	J	J	A	S	O	N	D
PLANT		▬	▬	▬							▬	▬
HARVEST						▬		▬				

FROM PLANTING TO HARVEST
1–3 years

VARIETIES
'Donau' – Austrian, heavy cropping

'Guincho Purple' – highly ornamental, finely cut deep purple leaves

'Golden Tower' – compact, upright, lime green foliage, good for pots

'Madonna' – vivid variegated cream and green leaves, compact, very attractive

ORIGINS
Hardy and adaptable European native

ALSO GROWS IN
Zones 1, 3, 4, 6, and 7

HARDINESS
Hardy

LIFE CYCLE
Large shrub

YIELD
★★/☆☆☆☆☆

EASY TO GROW
10/10

Highly ornamental, productive, and virtually indestructible – what's not to love? Size can be managed via pruning, and the flowers – the source of delicious drinks and preserves – are loved by insects. With huge berry clusters to boot, it's clear this plant is a winner.

Growing essentials

Planting Plant into soil dug over with garden compost or well-rotted manure. Tease out any spiralling roots, firm gently, and water well.

Pruning There's no need to prune – most varieties develop a beautiful vase-like trunk and canopy. If plants become too large, cut to shape, ideally in winter, though any time is possible.

Harvesting Harvest flowers as soon as they open – cut whole clusters and process within the day. Allow berries to ripen fully before harvesting to attain the best flavour – they freeze well.

Secrets of success

To keep compact, grow as multi-stemmed plants, rather than with a single trunk.

For container cultivation, choose a compact variety and grow in 50:50 loam and multi-purpose compost.

Harvest flowers whilst young, before the dusty yellow pollen is just released, for the best flavour.

Once harvested, leave flowers for a few hours for any insects to escape.

Berries are best eaten cooked as this breaks down any bitter flavour.

Look out for species *S. n.* subsp. *caerulea*, *S. n.* subsp. *canadensis*, and *S. racemosa* as fruiting forms.

Elder can be grown in many locations for the most exquisite floral harvests.

ZONE 5: PART SHADE

Mint
Mentha species

	J	F	M	A	M	J	J	A	S	O	N	D
SOW					▬							
TRANSPLANT						•					•	
HARVEST						▬▬▬▬▬▬▬▬▬						

FROM SOWING TO HARVEST
12–16 weeks

VARIETIES
M. spicata 'Tashkent' – vigorous spearmint, crinkled leaves

M. rotundifolia 'Bowles' – tall applemint, downy oval leaves

M. × *piperita* 'Black Mitcham' – peppermint, purple-flushed foliage

M. aquatica – water mint, happy in boggy soils and ponds

ORIGINS
A diverse group with worldwide distribution, revelling in areas of moist shade

ALSO GROWS IN
Zone 1, Zone 3, Zone 4, and Zone 6

HARDINESS
Hardy

LIFE CYCLE
Perennial

YIELD
★★★★★/☆☆☆☆☆

EASY TO GROW
9/10

With its unmistakable scent and unwavering determination to grow, mint is hugely popular across the globe. Whether you choose to add handfuls to a freshly made tabbouleh salad, marinate meats in it, or simply dot some sprigs into a refreshing summer cocktail, its consumption is bound to be memorable.

Growing essentials

Planting Generally grown from cuttings or divisions. Buy young plants and position in soil enriched with well-rotted organic matter. Tease out any spiralling roots first.

Mint spreads by rhizomes. To contain them, plant in a large container submerged into the ground. Leave a 2cm (¾in) "lip" above ground to deter rhizomes from escaping.

Harvesting Harvest when large enough. Remove individual leaves, or whole stems. Store leaves by drying, freezing, or making into mint jelly or other preserves.

Secrets of success

Seeds can be sown but germination is erratic – rhizomes and cuttings root readily.

Plants flower if stems are allowed to mature in summer – the blooms are loved by pollinators.

Mint is excellent for pots – make sure to keep plants well watered.

Divide congested plants when dormant using an old breadknife.

For fresh leaves in winter pot up sections in autumn and grow on warm windowsills.

There are many species – look also for *M. requienii, M. longifolia,* and *M. suaveolens*.

Spearmint (*M. spicata*) is an excellent partner in the kitchen for new potatoes, plus strawberries and raspberries, too.

ZONE 5: PART SHADE

Chilean guava
Ugni molinae

	J	F	M	A	M	J	J	A	S	O	N	D
PLANT				●					●			
HARVEST								▬▬				

FROM PLANTING TO HARVEST
1–3 years

VARIETIES
'Variegata' – attractive form, deep green leaves with mid-green edges

'Butterball' – mid-yellow young growth, fading to rich green

'Yanpow' – UK-bred, with a free-flowering and free-fruiting nature

'Flambeau' – leaves flushed with pink and edged with cream, on pink stems

ORIGINS
Cool and moist forests of Southwest America

ALSO GROWS IN
Zone 6, plus Zone 1 and Zone 3 if a little shade is available

HARDINESS
Hardy in all but the harshest of winters

LIFE CYCLE
Perennial

YIELD
★★★/☆☆☆☆☆

EASY TO GROW
8/10

Cultivated for its edible berries for centuries, their excellent flavour is laced with hints of pineapple and strawberry. Plants crop generously once established. Eat berries fresh, cooked, or dried – the bell-shaped blooms are very attractive, too.

Growing essentials

Planting Plant bushes in partial or full shade, into soil dug over with garden compost or well-rotted manure. Plants can be free-standing, or trained.

Pruning Plants need little pruning, and can be grown as a low hedge. Water during dry spells, and mulch every autumn or spring with bark chips or similar.

Harvesting Begin picking fruits late summer, through autumn. Eat fresh, and make gluts into jams as fruits are high in pectin, or freeze for storage.

Secrets of success

Plant in the warm soils of mid-autumn or mid-spring.

The bell-shaped blooms are scented and valuable to early emerging pollinators.

If space is limited loosely fan train plants against a structure.

Consider container growing. Keep roots cool and moist with thick-walled pots.

Young plants have limited hardiness but become more robust as they mature.

***U. montana* is a Mexican species** worth growing for its shiny black berries.

Chilean guavas make very useful contributions to the kitchen once plants establish.

Swiss chard and perpetual spinach
Beta vulgaris subsp. *vulgaris*

	J	F	M	A	M	J	J	A	S	O	N	D
SOW				▬	▬	▬	▬					
TRANSPLANT						▬						
HARVEST	▬	▬	▬	▬	▬	▬	▬	▬	▬	▬	▬	

FROM SOWING TO HARVEST
5–14 weeks

The sizeable leaves of 'Ruby' chard can grow soft and tender when shielded from strong sunshine.

Photogenic, productive, robust, rarely troubled by pests or disease – it's easy to see why these beets are revered. Myriad colours and nutritional benefits only increase the appeal. Simply braise and add leaves to curries, tarts, pasta, gratin, or soups.

Growing essentials

Sowing Sow 1cm (½in) deep in modules under cover in spring. Transplant once large enough, spacing 20cm (8in) apart, under cloches in frost-prone areas. Water and weed well until established.

Sow directly into prepared soil as spring warms. Excavate drills 3cm (1¼in) deep, sow seeds thinly, cover, and tamp down. Thin out to final spacings of 20cm (8in).

Harvesting Shear off baby leaf crops once foliage is 12–15cm (4¾–6in) tall. Water and feed well for another flush. Harvest larger plants when foliage develops – keeping a central rosette.

Secrets of success

Grow undercover as a cut-and-come-again leaf, for year-round harvests.

Water dry soils before sowing for good germination.

Eat any thinnings from direct sowings.

A midsummer sowing is useful for winter and early-spring pickings.

Harvest individual leaves from mature plants, twisting them off at the base.

Water in dry weather to deter bolting. Plants naturally run to seed in spring.

VARIETIES
Perpetual spinach – rich green, hardy, productive

'Lucullus' – glossy green sizeable leaves with thick white midribs, vigorous

'Bright Lights' – red, yellow, white, orange, pink-ribbed leaves

'Fantasy' – one of the tastiest red-ribbed chards, no bitter aftertaste

ORIGINS
Throughout Europe

ALSO GROWS IN
Zones 1, 3, 4, and 6

HARDINESS
Hardy, knocked back by extreme frosts

LIFE CYCLE
Biennial, grown as an annual

YIELD
★★★½/★★★★★

EASY TO GROW
8/10

ZONE 5: PART SHADE

Texsel greens
Brassica carinata

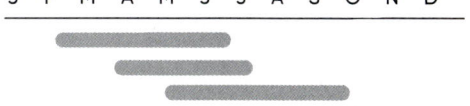

FROM SOWING TO HARVEST
8–12 weeks

VARIETIES
Sold simply as 'Texsel Greens'

ORIGINS
Widely grown in tropical and temperate countries due to its adaptable properties

ALSO GROWS IN
Zone 1 and Zone 3

HARDINESS
Hardy

LIFE CYCLE
Annual

YIELD
★★★½/☆☆☆☆

EASY TO GROW
7/10

Also known as African cabbage or Ethiopian kale, this brassica works as well in a sunny spot as it does in shade. Shade, though, brings a wonderful mild lushness to the cabbage-like leaves, and the flowering stems – which can be eaten like broccoli – remain tender and fibre-free.

Growing essentials

Sowing Sow for early crops undercover in modules. Sow one seed per cell, and transplant outside 30cm (12in) apart, under cloches in early to mid-spring. Protect plants from slugs and snails.

As soil warms in spring, sow directly into the soil. Make a drill 5cm (2in) deep, and sow seeds 1–2cm (½–¾in) apart. Thin out (and eat) emerging seedlings to 30cm (12in) apart.

Harvesting Harvest leaves as soon as plants are large enough – leave enough foliage so that plants can re-sprout for further pickings. Plants will ultimately run to flower as the season progresses.

Secrets of success

Protect from flea beetle, cabbage fly, and butterflies by cloaking in a fine mesh.

Avoid clubroot by adding garden lime to soils with an acidic pH.

Texsel greens are an adaptable crop, happy in sun but also thriving with some shade.

Achieve year-round harvests by sowing in August, for cropping under cover.

Regular harvests ensure continual tender, mild-tasting leaves.

When plants flower, leave them in place as they're edible and attract beneficial insects.

Let seedpods form at the end of the season, to collect and save your own seed.

Pak choi and relatives

Brassica rapa

	J F M A M J J A S O N D
SOW	▬▬▬
TRANSPLANT	▬▬▬▬
HARVEST	▬▬▬▬▬▬

FROM SOWING TO HARVEST
6–10 weeks

VARIETIES
Pak choi 'Red Choi' F1 – attractive, vigorous deep red rosettes

Pak choi 'Tai Sai' – large, productive, white-stemmed

Tatsoi 'Rozetto' F1 – squat, deep green rosettes, crinkled leaves, healthy and productive

Chinese cabbage 'Scansie' F1 – attractive, productive barrels, vivid pink leaves

ORIGINS
Asia, growing rapidly in warm, moist conditions where excess heat is avoided

ALSO GROWS IN
Zone 1 and Zone 3

HARDINESS
Not fully hardy – killed by frosts

LIFE CYCLE
Annual

YIELD
★★★★½☆☆☆☆☆

EASY TO GROW
7/10

Pak choi and tatsoi will give succulent, tender harvests when grown in a partly shaded site.

Thriving in cooler conditions with a moist root run, these leaves quickly bulk up. They're versatile, too, being useful as cut-and-come-again crops, baby vegetables, and mature plants for cooking. The thick, succulent leaves and crunchy midribs offer the perfect balance of textures.

Growing essentials

Sowing For an early crop, sow under cover in modules, one seed per cell, transplanting outside 30cm (12in) apart, under cloches early to mid-spring. Protect from slugs and snails.

As conditions warm, sow directly into soil. Excavate a drill 5cm (2in) deep, and sow seeds 1–2cm (½–¾in) apart. Thin (and eat) emerging seedlings to 30cm (12in) apart.

Harvesting Plants take 8–10 weeks to bulk up. Once sufficiently sized, harvest whole plants at the base using a sharp knife. If plants start to run to seed ("bolt"), eat the flower spikes and plants quickly.

Secrets of success

Pak choi or bok choy can quickly "bolt" in hot, dry areas.

Protect plants from slugs and snails, especially when establishing.

These crops work well in containers – be sure to keep compost well-watered.

You can grow these as microgreens, as well as to maturity.

Sowing early or late in the season, in cool weather, helps to deter bolting.

Plants mature quickly, especially in warmth – so sow regular batches for a succession of harvests.

Szechuan pepper
Zanthoxylum species

ZONE 5: PART SHADE

	J F M A M J J A S O N D
SOW	▬▬
PLANT	▬▬▬
HARVEST	▬▬

FROM SOWING TO HARVEST
3–5 years

VARIETIES
Z. simulans – Chinese Szechuan pepper, 4m (13ft) tall, yellow flowers, pink fruits

Z. piperitum – Japanese Szechuan pepper, 3m (10ft) tall, cream flowers, red fruits

Z. bungeanum – Nepalese Szechuan pepper, 3m (10ft) tall, vivid pink fruit clusters

Z. acanthopodium – lemon pepper, 2m (6½ft) tall, bright pink fruit tight to stems

ORIGINS
Cool, moist, mountainous regions of South and East Asia

ALSO GROWS IN
Zones, 1, 3, and 6

HARDINESS
Fully hardy

LIFE CYCLE
Perennial

YIELD
★★/☆☆☆☆

EASY TO GROW
8/10

Grown throughout Asia as an important spice for centuries. Eating the aromatic peppercorns will fill your mouth with a plethora of sensations: tingling, numbing, warming, electrifying – the flavour is beautifully intense. As a seasoning, it enlivens any dish.

Growing essentials

Sowing Sow seeds early spring into pots, under cover. Germination can be slow. Once large enough, transplant outside into final positions, water, and mulch well.

Pruning These shrubs don't require much pruning, and can be left to their own devices. Remove weak stems and cut lightly to shape, if required.

Harvesting Insignificant flowers appear early summer, followed by berry clusters. Harvest in autumn, dry off, and store in an airtight jar.

Secrets of success

Seed-raised plants can take years to crop – it's more common to buy plants.

Mix seeds with moist compost and then refrigerate in a bag for 1 month before sowing.

Individual plants do set fruit on their own, but will set more in groups.

The seeds themselves aren't eaten, but instead the surrounding seedcoats, which separate from the seeds as they dry.

Fresh tastes different to dried, and flavour varies between species.

Leaves and bark are edible and flavoursome – harvest these as a seasoning.

Szechuan pepper plants will crop readily in moist conditions with shade.

Ten other star performers

Welsh onion
Perennial Chinese clump-forming onion (*Allium fistulosum*) grown for the edible stems like a scallion, rather than bulbs. Reliably evergreen with fresh harvests year-round. White spherical flowerheads in summer, loved by bees.

Fuchsia
Hardy (*F. magellanica*), borderline hardy (*F. splendens* and *F. boliviana*), and tender (*F.* 'Coralle') species all fruit well. Plant hardy types in the ground and tender ones in pots to overwinter undercover. Fruits vary in pepperiness, flowers are highly ornamental.

Mashua
Highly ornamental climber (*Tropaeolum tuberosum*) yielding knobbly, protein-rich tubers. Though borderline hardy, the tuber's spicy flavour is mellowed by light frosts. The nodding, yellow-orange flowers of 'Ken Aslet' are especially attractive.

Coriander
Often running to seed in full sun. This hardy annual Mediterranean herb (*Coriandrum sativum*) produces abundant foliage if given a little shade. It quickly bulks up from seed and freezes well. Roots and seeds are also edible.

Common brighteyes
Mediterranean perennial *Reichardia picroides* offers a low-maintenance alternative salad leaf. Happy in sun or part shade. Pretty yellow aster-like flowers all summer, pollinator-loved. Self-seeds, seems slug resilient, too.

Coriander (below left) is slower to flower in shade; brighteyes (below right) shows good tolerance to shade and slug damage.

ZONE 5: PART SHADE   137

Lovage

A European hardy perennial herb, *Levisticum officinale* bears strongly cut leaves with celery-like flavour. Fresh spring growth is especially palatable – mellow the intensity of mature foliage by forcing under a large pot. Attractive flower umbels are a wildlife magnet.

Garlic cress

A brassica family member, found throughout Eastern Europe. Perennial woodland species (*Peltaria alliacea*) growing to 1m (3ft) high. Young leaves have a pleasant and mild garlic flavour that intensifies with age. Attractive white flowers and disc-like seedpods.

Serviceberry

Various North American species of *Amelanchier* (*A. alnifolia*, *A.* × *lamarckii*, *A. confusa*, *A. stolonifera*). Abundant, ornamental, white starry flowers in spring, followed by clusters of berries in various colours, eat raw or cooked.

Komatsuna

Brassica rapa, also known as Japanese greens, grown for oval, peppery leaves. Quick to bulk up. Grow as baby leaf, or to maturity. Plants are hardy and thrive in cooler months. Repeat sow for successional harvests.

Nasturtiums

In shades of orange, red and yellow, the vivid flowers of *Tropaeolum majus* bring a splash of colour to a shadier area. Blooms are edible, as are the leaves and seeds. All have a hot, peppery flavour. Best grown as a tender annual. Will happily self-seed once established.

Komatsuna (top) quickly bulks up in cool weather; nasturtiums (above) make good ground cover in shade, offering harvests of leaves, flowers, and seeds.

Project:
Creating a beautiful wild berry patch

Where would we be without strawberries, apples, and raspberries? Well, we'd be foraging, and that's exactly what you can do in your shrub borders. Happy in part shade, the berries listed here offer a refreshing alternative to our usual fruits. By growing them we utilize borders for edibles and bolster supplies for desserts, preserves, and sauces.

There are dozens of fruiting plants that can be grown in shaded gardens (see suitable berries list, opposite below), from health-boosting barberries and goji berries, to flavoursome Chilean guavas and gaultherias. Those chosen here are hardy perennials to ensure that your berry bed will provide you with fruits for many years to come.

With such a range to choose from, remember to check the ultimate height and spread of your chosen berries before planting them. All are happy in part shade. This lack of sunshine gives some species a naturally sharp flavour, so experiment with these to create jams, jellies, or other sugar-boosted preserves. When growing berries for foraging it also pays to accept that birds may eat a few (think of this as a compliment!). While netting to protect the fruits is possible, a "live and let live" mindset might prevent your garden from looking like a fortress.

(Above from left to right) *Berberis vulgaris*, *Berberis darwinii*, *Gaultheria mucronata*, and *Ugni molinae* would all provide harvests in a shady border.

ZONE 5: **PART SHADE**

You will need

Fruiting plants
Fork or spade
Slow-release fertilizer
Organic mulch

Steps

1 Position the plants in their pots in your chosen bed, moving them around until you are happy. Photograph your final choice.

2 Remove the plants and dig over the bed, working fertilizer into the soil. Excavate planting holes for each fruit, slightly wider than the rootball.

3 Knock the plants out of their pots, gently teasing out any spiralling roots, and plant.

4 Plant, water in well, and lay a mulch of 4–5cm (1½–2in) deep, over the whole area.

Suitable berries:

Temu (*Temu cruckshanksii*)*
Chilean myrtle (*Luma apiculata*)*
Jostaberry (*Ribes* × nidigrolaria)
Worcesterberry (*Ribes divaricatum*)
Cathay quince (*Chaenomeles cathayensis*)
Chilean wineberry (*Aristotelia chilensis*)*
Japanese plum yew (*Cephalotaxus harringtonia*)
Shallon (*Gaultheria shallon*)
Autumn olive (*Elaeagnus umbellata*)
Hardy fuchsia (*Fuchsia magellanica*)
Oregon grape (*Mahonia aquifolium*)
Chilean guava (*Ugni molinae*)*
Honeyberry (*Lonicera caerulea*)
Darwin's Barberry (*Berberis darwinii*)
Goji berry (*Lycium barbarum*)
Black chokeberry (*Aronia melanocarpa*)

*hardy to -5°C (23°F)

Gardeners in possession of Zone 6 conditions could mistakenly believe this area promises scant yields, but nothing could be further from the truth. Many crops thrive in these cool, sheltered conditions, where delicate leaves are allowed to unfurl, unbuffeted and unscorched by weather extremes. Thick, plump stems of rhubarb and celery can be enjoyed, as can tender shoots of hosta, and shuttlecock ferns. Lush harvests of winter purslane, mizuna, minutina, and sorrel will add a plethora of flavours to salads. Bulky, shade-loving greens like Caucasian spinach and bulbous nettle can be the backdrop for herbs like parsley, heart's ease, and chervil. Blackcurrants, prickly heath, and chokeberries add fruit to this exciting mix.

Zone 6
Shady and wet

Celeriac and celery
Apium graveolens

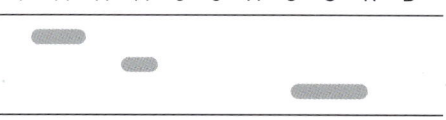

FROM SOWING TO HARVEST
28–32 weeks

Unique, characteristic flavours make up for slow growth in these two crops. Whether you dream of fireside winter teas of cheese, crackers, and celery, or warming celeriac casseroles, grills, and soups, your cool-season recipes will benefit from your patience.

Growing essentials

Sowing Sow seeds shallowly, into pots or trays of seed compost. Germination can take 3–4 weeks so be patient. Keep compost neither dry nor waterlogged until seedlings appear.

Planting Transplant seedlings outside, 30cm (12in) apart, into soil with ample organic matter added. Keep well watered and protect from slugs in rainy periods.

Harvesting Weed well, and apply a balanced liquid feed periodically throughout summer. Harvest when large enough in autumn.

Secrets of success

Plant out seedlings before they become rootbound, to deter growth checks.

Mulch around roots with organic matter to suppress weeds and conserve moisture.

Alleviate leaf miner by immediately removing leaves showing symptoms.

Remove celeriac outer leaves in summer to expose the swollen base.

Lift celeriac before frosts, remove leaves and roots and store in boxes of damp sand.

Protect celery throughout winter with cloches cloaked in old sheets and blankets.

VARIETIES
Celeriac 'Monarch' – healthy, rich green foliage, pure white flesh

Celeriac 'Prinz' – good-sized harvests, bolt resistant

Celery 'Lathom Self-Blanching' – reliable, good-sized, robust

Celery 'Loretta' – high quality, little stringiness, British-bred

ORIGINS
The Mediterranean basin

ALSO GROWS IN
Zone 2, Zone 3, Zone 4, and Zone 5 with ample moisture available

HARDINESS
Hardy in all but harsh winters

LIFE CYCLE
Biennial, grown as an annual

YIELD
★★★/☆☆☆☆

EASY TO GROW
6/10

Given a steady supply of moisture, both celeriac and celery will quietly bulk up come the season's end.

ZONE 6: SHADY AND WET

Rhubarb
Rheum × hybridum

	J	F	M	A	M	J	J	A	S	O	N	D
PLANT				▬	▬						▬	▬
HARVEST				▬	▬	▬	▬					

FROM PLANTING TO HARVEST
1–2 years

VARIETIES

'Livingstone' – compact, dormant for harvesting in early autumn

'Timperley Early' – robust, early cropping, thick red-green stems

'Raspberry Red' – healthy, sweet red stems, no need to force

'Champagne' – vigorous, reliable, early/mid-season, for forcing

ORIGINS
Europe and Scandinavia, giving rise to robust plants that enjoy moisture

ALSO GROWS IN
Zones 1, 3, 4, and 5

HARDINESS
Hardy

LIFE CYCLE
Perennial

YIELD
★★★★★/☆☆☆☆☆

EASY TO GROW
10/10

Ridiculously generous and oh-so-easy to grow. Rhubarb is popular with novice and experienced gardeners alike. The thick, sweet-sour stems make excellent harvests, and forcing transforms them into something uniquely tender, delicate, and wonderful.

Growing essentials

Planting Best planted dormant in winter, when they're available bare root. Potted plants are available in spring. Enrich soil with ample well-rotted organic matter.

Aftercare Resist the temptation to harvest the first year – allow plants to bulk up. Weed well until established and mulch annually with well-rotted organic matter.

Harvesting Harvest once full sized. Take only a few stems the second year, then a third once established. Allow forced plants to recover by not forcing the following year.

Secrets of success

To plant bare root position the growing point at or fractionally above the soil.

Sow seeds in spring. Seedlings may be variable and slow to bulk up.

Remove flower spikes when they appear to focus energy into the leaves.

Divide crowns every 4–6 years, in late autumn. Slice the fleshy roots into clumps of 3–4 buds.

To force rhubarb, cover with a large upturned pot or forcing jar. Lay organic slug controls.

Harvest the red/pink stems by sliding your thumb and index finger to the base of the stem, then twisting it off.

The sizeable leaves of rhubarb appreciate a position that offers some shade.

Winter purslane
Claytonia perfoliata

	J	F	M	A	M	J	J	A	S	O	N	D
SOW				▬	▬			▬	▬			
TRANSPLANT					▬							
HARVEST				▬	▬	▬	▬	▬			▬	▬

FROM SOWING TO HARVEST
10–14 weeks

VARIETIES
Generally sold as the basic species

ORIGINS
Found naturally along woodland margins in North America, thriving in cool, moist autumn and spring

ALSO GROWS IN
Zone 3, Zone 4, and Zone 5

HARDINESS
Hardy

LIFE CYCLE
Annual

YIELD
★★★★/☆☆☆☆

EASY TO GROW
8/10

While there's an abundance of tasty leaves to satisfy our summer salad appetite, the winter range is more limited. Thankfully this quick-growing, very hardy crop, also known as miner's lettuce, delivers abundant harvests of fleshy, mild-tasting pickings.

Growing essentials

Sowing Sow seeds shallowly in small pots or trays, under cover, in spring. Once seedlings are large enough to handle, transplant individually into large modules and grow on under cover.

Planting Transplant outside once roots are well established, at 15cm (6in) spacings. Lay organic slug controls whilst plants bulk up, and keep weed-free.

Harvesting Pick as soon as leaves are large enough. Either snip off individual leaves as needed – leaving the growing point intact allows multiple harvests – or harvest the whole rosette.

Secrets of success

Sow seeds late summer or early autumn, to overwinter and provide winter pickings.

If allowed to flower and set seed, winter purslane will happily colonize your garden.

Plants are incredibly winter hardy, surviving happily at -20°C (-4°F), making them invaluable for harsh zones.

To encourage soft, lush growth in winter, grow under cloches or in cold frames.

Plants will naturally self-seed and die in summer. New seedlings appear in autumn.

Look for seeds or plants of pink purslane (*C. sibirica*) as another winter salad leaf.

Winter purslane is a great shade-tolerant salad leaf for cool-season harvests.

ZONE 6: SHADY AND WET

Blackcurrants
Ribes nigrum

	J	F	M	A	M	J	J	A	S	O	N	D
PLANT			●								●	
HARVEST							●					

FROM PLANTING TO HARVEST
1–3 years

VARIETIES

'Ben Connan' – compact, early season, excellent flavour and yield

'Ben Sarek' – mid-season, for cooking or eating fresh, compact

'Big Ben' – upright plants, grown for large berries, dual purpose, early season

'Ben Lomond' – vigorous, heavy crops, ripening late season

ORIGINS
Widely distributed in the Northern Hemisphere

ALSO GROWS IN
Zone 1, Zone 3, Zone 4, and Zone 5

HARDINESS
Hardy

LIFE CYCLE
Perennial

YIELD
★★★★☆☆☆☆☆☆

EASY TO GROW
8/10

There are few fruits that provide such a concentrated flavour as blackcurrants. Their unmistakable taste has cemented their cordial as a household name, and overly generous harvests year on year means you should definitely afford them space on your plot.

Growing essentials

Planting Bushes best planted when dormant. Add ample well-rotted bulky organic matter to weed-free soil before planting – blackcurrants are especially hungry plants.

Pruning Blackcurrants fruit predominantly on younger wood – 1-, 2-, and 3-year-old stems. Young plants therefore need little pruning as they bulk up. Remove a proportion of older wood annually, on mature bushes.

Harvesting Leave individual fruit clusters, or "strigs", to ripen fully, before carefully picking the stalks from the plant. Sitting alongside bushes can save your back and knees when making sizeable harvests.

Secrets of success

Cut back young bushes hard on planting to improve root establishment.

Renovate unproductive plants by cutting all stems to ground level in winter.

Prune bushes in winter. Alternatively, remove whole older stems when harvesting for berry access.

If you have a bed that remains moist in summer, blackcurrants will happily grow there and crop abundantly.

While these plants are hardy, subzero temperatures during flowering can limit yields, so avoid planting in frost pockets.

Feed blackcurrants annually, especially on light soils. Apply both organic-based potash and nitrogen feeds.

Bird protection is essential – whilst fruits ripen, erect a wire cage over bushes.

ZONE 6: SHADY AND WET

Prickly heath
Gaultheria mucronata

	J	F	M	A	M	J	J	A	S	O	N	D
PLANT				▬						▬		
HARVEST								▬▬▬				

FROM PLANTING TO HARVEST
1-3 years

VARIETIES
'Bell's Seedling' – medium-sized shrub, vivid, deep pink fruits

'Crimsonia' – dusky pink fruits, moderate vigour

'Mulberry Wine' – healthy plants, mid-pink fruits deepening with age

'Snow White' – compact form growing in pots, pure white fruits

ORIGINS
Mountainous Southwest America, in cool, moist, well-drained soils

ALSO GROWS IN
Zone 4 and Zone 5 – if sufficiently moist

HARDINESS
Very hardy

LIFE CYCLE
Perennial

YIELD
★★★/☆☆☆☆

EASY TO GROW
8/10

These sprawling evergreen shrubs offer highly attractive berries in a range of colours, and low-maintenance, trouble-free growing. The berries have a pleasantly spongy texture – a refreshing snack straight off the bush. The leaves are also highly aromatic when crushed.

Growing essentials

Planting Plant in spring or autumn into well-drained yet moisture-retentive soil enriched with acidic organic matter. Water well and add a thick layer of mulch.

Pruning Bushes need very little pruning, other than a light shape when unbalanced. Acidic conditions preferred so mulch annually each spring with ericaceous compost.

Harvesting Harvest berries once fully mature, picking over bushes for a month or so to enjoy fruits at their best. Best eaten fresh though gluts can be frozen.

Prickly heath offers generous harvests in Zone 6, on low-maintenance plants.

Secrets of success

Usually plants are bought but stock can be bulked up with rooted suckers or seed.

Bell-shaped nodding flowers appear early summer – much loved by pollinators.

Shows strong honey fungus resistance – useful if this is troublesome locally.

Groups of plants make excellent low-maintenance groundcover for busy gardeners.

Plants are dioecious so a male form, such as *G.* 'Mascula' must be nearby for berry production.

Other species such as *G. procumbens* and *G. shallon* are also worth growing as edibles.

Parsley
Petroselinum crispum

	J	F	M	A	M	J	J	A	S	O	N	D
SOW			▬	▬								
TRANSPLANT				▬	▬							
HARVEST	▬	▬	▬	▬	▬	▬	▬	▬	▬	▬	▬	▬

FROM SOWING TO HARVEST
16–24 weeks

A go-to herb for many due to its abundant harvests and versatility in the kitchen. Just one or two parsley plants makes a household self-sufficient all year, leading to flavoursome casseroles, tabboulehs, pestos, breads, and beyond.

Growing essentials

Sowing Sow seeds, shallowly, into pots or trays of seed compost. Germination can take 3–4 weeks so be patient. Alternatively, sow seeds directly into well-prepared, warm spring soil.

Planting Transplant seedlings out, 30cm (12in) apart – or into large pots. Water well (parsley dislikes drying out) and feed with balanced liquid feed through the season.

Harvesting Harvest individual stems from plants once established. Remove yellowing, older leaves to keep tidy and productive.

Secrets of success

Keep seed trays just moist until seedlings emerge, to avoid drying out or waterlogging.

Plants thrive in cool autumns, overwintering to yield well into spring.

If root fly is troublesome, rotate fresh sowings or plantings to new beds.

Store gluts by freezing in bags – crunch the bag up to chop leaves.

Curly parsley tastes mild and delicate, flat-leaf has a stronger flavour.

For an indoor crop, grow mature plants in pots or as seedling microleaves.

VARIETIES
'Moss Curled 2' – beautifully curled green leaves, strong, stocky

'Giant of Italy' – flat-leaf form, vigorous, rich green

'Hamburg' – grown for fleshy white roots, foliage also edible

'Curlina' – compact moss-type, ideal for containers

ORIGINS
Mediterranean basin, optimum longevity when in moist, temperate regions

ALSO GROWS IN
Zones 1, 3, 4, and 5

HARDINESS
Fully hardy

LIFE CYCLE
Biennial, often grown as an annual

YIELD
★★★★/☆☆☆☆

EASY TO GROW
8/10

Both moss- and flat-leaf parsleys can give a long season of pickings in shadier, moist locations.

Hosta

Hosta species

	J	F	M	A	M	J	J	A	S	O	N	D
PLANT				●						●		
HARVEST				▬	▬							

FROM PLANTING TO HARVEST
1–3 years

VARIETIES

H. longipes – mid-green leaves, purple-pink summer flowers

H. montana **'Aureomarginata'** – vigorous, gold-margined leaves

H. fortunei var. *albopicta* – strong, cream leaves edged with streaks of green

H. sieboldiana var. *elegans* – striking, large blue-grey leaves, vigorous

ORIGINS
A diverse genus, originating from moist woodland soils and gentle shade in Asia

ALSO GROWS IN
Zone 5

HARDINESS
Fully hardy

LIFE CYCLE
Perennial

YIELD
★★½/☆☆☆☆

EASY TO GROW
7/10

"Hostons" of mature plants can be harvested and eaten, then subsequent shoots can be left to unfurl.

Well known and much loved as an ornamental, all parts of the hosta are edible. Allowing clumps to build up gives you decent harvests of the "hostons" (curled up shoots) in spring. Plants can then be left to re-sprout so that gardeners can appreciate their stunning leaves and flowers.

Growing essentials

Planting Buy young and plant in a soil enriched with well-rotted organic matter. Tease out spiralling roots first. Water well to establish and protect from slugs.

Aftercare Mulch plants annually in spring with garden compost or well-rotted manure. Cut back faded foliage in winter. Hostas will gently bulk up year on year. Harvest once clumps are sizeable.

Harvesting Harvest shoots when plants are large enough. Snap off hostons as they emerge. Unfurled leaves are edible, but a little more fibrous once mature.

Secrets of success

If slugs are problematic carry out nightly patrols with a torch to reduce.

Only harvest the first flush of shoots, then leave plants to re-sprout and recover.

Container cultivation is popular – fill a large pot with loam-based compost.

Divide congested clumps in autumn or early spring, to keep productive.

Remove outer leaf scales from hostons, as these taste fibrous and bitter.

Flowers and young flowering shoots are edible as a midsummer harvest.

ZONE 6: SHADY AND WET

Dog's tooth violet

Erythronium species

	J F M A M J J A S O N D
PLANT	• •
HARVEST	━━━━

FROM PLANTING TO HARVEST
2–4 years

VARIETIES
E. dens-canis – sturdy and strong, mottled leaves, mauve flowers

E. americanum – needs time to bulk up, mottled leaves, cream flowers

E. japonicum – gently mottled leaves, dusky pink flowers

E. grandiflorum – vigorous, rich yellow blooms above green leaves

ORIGINS
Throughout the Northern Hemisphere

ALSO GROWS IN
Zone 3, Zone 4, and Zone 5

HARDINESS
Fully hardy

LIFE CYCLE
Bulbous perennial

YIELD
★★/☆☆☆☆

EASY TO GROW
8/10

The flowers of erythroniums make these highly ornamental plants – the edible bulbs are a tasty bonus.

These highly attractive and desirable plants grace many a woodland border with their two-tone foliage and nodding reflexed flowers. Once established, clumps of edible species can be uprooted in summer and early autumn to harvest the starchy bulbs. Shaped like a dog's tooth, they are crunchy, sweet, and very pleasant.

Growing essentials

Planting Buy young potted plants in spring, or dormant bulbs in autumn. Plant in soil enriched with well-rotted organic matter. Water in well, to establish, lay an organic mulch and protect from slugs.

Aftercare Mulch annually in early spring with garden compost or well-rotted manure. Do not deadhead – allow plants to self-seed. Cut back faded foliage in summer. Harvest once clumps have bulked up.

Harvesting Lift and eat the bulbs in spring, when they have a crisp and juicy texture. If you want to enjoy the flowers, delay harvesting until the plants' leaves die down in early summer.

Secrets of success

Plant bulbs deeply – at least 15cm (6in) deep to deter drying out.

Leaves are edible but harvesting them severely reduces plant vigour.

To bulk up clumps, carefully lift and divide immediately after flowering.

Store harvested bulbs in damp sand, to deter excessive withering.

The bulb's flavour is initially mild, but sweetens as it matures in storage.

The starchy bulbs can be dried and made into flour for a thickening agent.

Mizuna and mibuna

Brassica rapa

	J	F	M	A	M	J	J	A	S	O	N	D
SOW			▬	▬	▬	▬	▬	▬				
TRANSPLANT				▬	▬	▬	▬	▬	▬			
HARVEST					▬	▬	▬	▬	▬	▬	▬	

FROM SOWING TO HARVEST
6–10 weeks

VARIETIES
Mibuna – generally available as the basic crop, rosettes of spoon-like, deep green leaves

Mizuna – frequently sold simply as "mizuna", clumps of rich green, deeply-cut leaves

Mizuna 'Red Empire' – attractive, gently-lobed red leaves

Mizuna 'Kyoto' – vigorous, upright clumps, serrated leaves

ORIGINS
Widely grown, thriving in cool conditions

ALSO GROWS IN
Zone 1, Zone 3, Zone 4, and Zone 5

HARDINESS
Hardy

LIFE CYCLE
Annual

YIELD
★★★★½/☆☆☆☆☆

EASY TO GROW
9/10

These two fast-growing annual brassicas are excellent edibles for filling the otherwise lean harvest periods of early spring and late autumn. The pleasantly peppery leaves are produced with huge generosity – only a few plants are needed for most households. Their compact and productive nature makes them a must-grow crop.

Both mibuna (top) and mizuna, (here, 'Red Empire', bottom) enjoy spring and autumn conditions.

Growing essentials

Sowing Sow seeds 1cm (½in) deep, 2–3 seeds per cell, in modules, under cover in spring. Transplant these outside once large enough, spacing plants 20cm (8in) apart – cloches will speed up growth. Water well until established.

As spring warms, sow directly outside, in shade, into prepared soil. Excavate drills 3cm (1¼in) deep, sow seeds thinly, cover, tamp down. Thin out seedlings to final spacings of 20cm (8in).

Harvesting Shear off plants grown as a baby leaf crop once foliage is 12–15cm (4¾–6in) tall. Water and feed well for another flush. Harvest larger plants as and when foliage develops – always leaving a central rosette of leaves.

Secrets of success

Both crops are less productive in summer heat – focus on spring and autumn crops.

Being brassicas, it's wise to protect from cabbage root fly with fine mesh.

Winter crops are easily gained under cloches or in polytunnels.

High temperatures cause extreme pepperiness – shade is preferable.

Grow crops as a microleaf or cut-and-come-again crop on windowsills in winter.

Their compact nature makes them excellent for container-growing.

ZONE 6: SHADY AND WET

Minutina
Plantago coronopus

	J F M A M J J A S O N D
SOW	▬▬▬
TRANSPLANT	▬▬
HARVEST	▬▬▬▬▬▬▬▬▬▬▬▬

FROM SOWING TO HARVEST
12–16 weeks

VARIETIES
Available as the basic species

ORIGINS
Widely distributed, adaptable species

ALSO GROWS IN
Zones 1, 2, 3, 4, and 5

HARDINESS
Fully hardy

LIFE CYCLE
Annual or short-lived perennial

YIELD
★★★☆/☆☆☆☆

EASY TO GROW
9/10

Minutina, buck's-horn plantain, or *erba stella* in Italy where it is incredibly popular, is a highly productive hardy crop used as a salad leaf and a steamed green vegetable. It originates from coastal areas but is very versatile, growing in semi- or deeper shade.

Growing essentials

Sowing Sow seeds for early crops, under cover in modules. Sow 2–3 seeds per cell, and transplant outside 30cm (12in) apart, under cloches in early to mid-spring. Protect from slugs and snails.

As soil conditions warm in spring, sow directly into the soil. Excavate a drill 5cm (2in) deep, and sow seeds 1–2cm (½–¾in) apart. Thin (and eat) emerging seedlings to 30cm (12in) apart.

Harvesting Harvest leaves when plants are large enough – leave sufficient foliage so they can re-sprout for further pickings. Plants ultimately run up to flower, which attracts beneficial insects.

Both the leaves and young flower spikes of minutina can be eaten.

Secrets of success

Once plants mature, flower spikes appear – eat these while they are young and in bud.

Minutina will self-seed, so don't remove the flower spikes if you desire this.

Year-round leaf production is easily achieved growing in a polytunnel.

Plants are perennial, but sow fresh stock regularly as their vigour naturally declines.

For tender leaves harvest regularly, as this encourages vigorous fresh growth.

Leaves become tough and bitter in strong sunshine – shade is preferable.

Ten other star performers

Caucasian spinach
Vigorous twining perennial vine (*Hablitzia tamnoides*) native to Asian woodland ravines, thriving in Zone 6. Incredibly hardy, growing to 3m (10ft) tall in one season when mature. Readily raised from seed, yielding for many years once established.

Black chokeberry
Resident of damp American woods, *Aronia melanocarpa* produces generous clusters of shiny black berries. Tart raw, excellent cooked with sugar. Jam-makers prick up your ears as they're high in pectin. Best on acid soils.

Shuttlecock fern
Eat the young emerging fronds of this hardy Northern Hemisphere species (*Matteuccia struthiopteris*) as "fiddleheads". Harvest while small and tender, and cook (not edible raw). To avoid overcropping plants, only take one-third of the fronds.

Heart's ease
This variable short-lived European perennial, known botanically as *Viola tricolor,* brings colour to shaded areas and a pretty harvest. While the leaves and young buds are edible, it's the flowers that are worth gathering, to add to salads and garnish bakes.

(Left) The edible flowers of heart's ease bring additional colour to a cool, shady border.

(Below right) Chervil is a dainty leafy herb that will gently naturalize in shady, damp areas.

(Below) Red-veined sorrel (*Rumex sanguineus*) is a useful perennial for shadier plots, filling May "hungry gap".

ZONE 6: SHADY AND WET

Shuttlecock ferns are highly ornamental plants that bear edible young fronds.

A cool, shady position can be occupied by bulbous nettles to give you a nutritious leafy green.

Chervil

Anthriscus cerefolium, an annual herb with feathery foliage, brings a uniquely delicious warmth to savoury dishes. Quickly raised from direct-sown seed. Though hailing from Southeast Europe, it is fully hardy. Produces frothy white flowers in summer, self-seeding happily.

Bulbous nettle

Perennial nettle (*Laportea bulbifera*) native to Japanese forest valleys, thriving in moist shade. Like its better-known relative, *Urtica dioica*, the stinging hairs on the leaves are neutralized by cooking. Hardy, prolific, propagated via aerial bulbils or division.

Mitsuba

Or Japanese parsley, a borderline hardy perennial (*Cryptotaenia japonica*) easily raised from seed. Quickly bulks up to form a large mound of foliage. Flavour reminiscent of chervil and celery. Blanch mature plants, and use seedlings as a microleaf.

Sorrels

Two genera, *Oxalis* and *Rumex*, provide excellent edible sorrels. Leaves of *O. acetosella* and *O. oregana* (both fully hardy), plus *O. articulata* (borderline hardy), along with *R. acetosa*, *R. acetosella*, *R. sanguineus*, and *R. scutatus* give a pleasant lemony flavour. *Oxalis* bear pretty flowers.

Wasabi

Also called Japanese horseradish, *Eutrema japonicum*, is a fleshy rhizome renowned for fiery flavour. Found by woodland streams, growing slowly. Give young transplants a few years to bulk up. Grate, then eat immediately for the best flavour.

Chinese artichoke

Incredibly hardy and thirsty. *Stachys affinis* produces spikes of lilac flowers if grown in sun. Happy in shade, though not so free-flowering, tubers bulk up well. Store in just damp sand once lifted in late autumn.

Mushroom cultivation requires moisture and shade – hello Zone 6! Impregnating logs with mushroom dowels really boosts your grow-your-own kudos, who isn't impressed by home-grown oyster or shiitake mushrooms? The process is simple to start, then it just needs time.

Project:
Growing your own mushroom "loggery"

Most of us are familiar with button mushrooms, but gourmet species such as oyster, shiitake, lion's mane, and nameko can be grown on logs outside. The shady, damp conditions are ideal for fungal growth and the logs, once rotted, make excellent habitats for rare or beneficial predatory creatures, such as stag beetles, devil's coach horse beetles, and toads.

Logs and spawn

If you have an established garden you may find logs suitable for mushroom growth on your plot. A local tree surgeon is an excellent source of these timbers if not. Freshly felled hardwoods like oak, beech, and ash are preferred. Look for non-decayed wood, without wounds in the bark, otherwise your log may already be home to unwanted fungi. Source spawn from mushroom growing specialists – it usually comes in the form of wooden dowels. Use it as soon as it arrives, ensuring you have logs ready. Store spawn in the fridge for a few weeks, if not. The process of impregnating logs with spawn is simple, with one log frequently giving repeat flushes of mushrooms.

ZONE 6: **SHADY AND WET**

You will need

Drill
8mm (size O) drill bit
Logs
Mushroom spawn (dowels)
Hammer
Lighter
Wax, to plug

Steps

1 Drill holes in your logs, 5cm (2in) deep and spaced 10cm (4in) apart. Before drilling take precautions like tying back loose hair, securing loose clothing, and eye protection.

2 Carefully push individual dowels into your log, driving them home with a hammer. Use one mushroom species per log, rather than mixing them, otherwise vigorous species will out-compete others.

3 Seal each hole with melted wax (beeswax is ideal) to deter other fungi from entering – take care when melting wax. Your log is now inoculated with your chosen mushrooms.

4 Position your logs in your shady, damp location. Keep the logs upright, burying the bottom 10cm (4in) in the soil to ensure they're stable. Mushrooms should appear in 9–12 months.

Vigorous root systems and an unwavering zeal for life bring many exciting edibles to this outwardly unwelcoming zone – and the addition of mulch also helps! Forest fruits are surprisingly abundant here. Look to gooseberries, currants, blackberries, and their many relatives as the backbone of this zone, with barberries, alpine strawberries, honeyberries, and Oregon grapes bolstering harvests. Cobnuts, too, crop profusely in their native understorey setting. Robust woodland leaves, like wild garlic, garlic mustard, and nettle will deliver trugs full of greens, and deep-rooted patience dock, bamboo, udo, and horseradish can extract all the nutrition they need from this seemingly impoverished soil.

Zone 7
Shady and dry

ZONE 7: SHADY AND DRY

Gooseberries

Ribes uva-crispa

	J	F	M	A	M	J	J	A	S	O	N	D
PLANT	▬	▬									▬	▬
HARVEST						▬	▬	▬				

FROM PLANTING TO HARVEST
1–3 years

VARIETIES
'Invicta' – vigorous, mildew resistant, very productive, green culinary fruits

'Hinnonmäki Röd' – tasty, smooth-skinned red dessert fruits, healthy plants

'Early Sulphur' – delicious, yellow-fruited dessert variety, thorny

'Leveller' – flavoursome, green-fruited dessert type, large attractive berries

ORIGINS
Rocky, woodland, mountainous regions of Europe

ALSO GROWS IN
Zones 1, 3, 4, 5, and 6

HARDINESS
Fully hardy

LIFE CYCLE
Perennial

YIELD
★★★★/☆☆☆☆

EASY TO GROW
8/10

Gooseberries grow in a wide range of zones, but in particular will crop in dry shade. Many of us are familiar with the green, tart cooking fruits – but I invite you to indulge in the dessert types as they bring a sweeter, pop-in-the-mouth harvest.

Growing essentials

Planting Plant bushes when dormant, enriching weed-free soil with well-rotted organic matter, and mulching. This helps build a robust, vigorous root system.

Pruning Prune bushes in winter: create a balanced framework of main stems on young plants. On mature bushes, remove congested growth and a proportion of very old stems.

Harvesting Harvest fruit once large and sweet, midsummer onwards. Pick over bushes three or four times. Gluts freeze well for cooking later.

Secrets of success

Train as fans or cordons for space efficiency in smaller gardens – prune these also in summer.

Buds are produced on mature wood – aim to build a framework of established stems.

Take hardwood cuttings from healthy stems when pruning in winter.

Caging ensures plants aren't raided by birds – avoid netting as birds can become entangled.

Culinary varieties are best for deep shade. Dessert types enjoy a little sun.

For especially large fruits – for exhibition for example – thin out fruitlets in late April or early May.

Red-fruited gooseberry 'Hinnonmäki Röd' grows well in a cooler, shadier spot.

Red, pink, and whitecurrants

Ribes rubrum

	J F M A M J J A S O N D
PLANT	▬▬ ▬▬
HARVEST	▬

FROM PLANTING TO HARVEST
1–3 years

Known for eye-candy good looks and abundant yields, these currants survive with ease where many edibles struggle. Mature bushes deliver significant crops, and the characteristically tart flavour means berries don't seek out much sun.

Growing essentials

Planting Bushes best planted when dormant. Add ample well-rotted, bulky organic matter to a weed-free plot before planting, and mulch well to lock in soil moisture.

Training These plants bear a good amount of fruit on older wood, so train permanent stems against shady walls or fences. Alternatively allow to develop into open bushes.

Harvesting Let fruit clusters – "strigs" – ripen fully, before carefully picking stalks from plants. Ripening can be gradual, be patient to ensure the best flavour.

Secrets of success

Prune out congested growth in winter, and cut back excess growth in summer.

Smaller plots can grow these currants as cordons, spur-prune back to keep compact.

(Top) Whitecurrant 'White Versailles' may look delicate but bushes are surprisingly robust. (Above) Redcurrant 'Rovada' will readily yield sizeable harvests against a shady north-facing wall.

Take hardwood cuttings in winter when pruning healthy plants.

Erect wire mesh over plants as soon as fruits begin to colour, for bird protection.

Periodically remove older stems on mature plants, to encourage the renewal of vigorous growth.

Gluts freeze well for cooking later – these fruits make excellent jellies and jams.

VARIETIES

Redcurrant 'Stanza' – mid-season variety, high yields of large berries

Redcurrant 'Rovada' – vigorous bushes cropping heavily late in season

Pinkcurrant 'Gloire de Sablons' – beautiful pink berries borne midsummer

Whitecurrant 'White Versailles' – virtually translucent white fruits

ORIGINS
Woodland margins of West Europe, where plants enjoy gentle shade and a cool root run

ALSO GROWS IN
Zones 1, 3, 4, 5, and 6

HARDINESS
Fully hardy

LIFE CYCLE
Perennial

YIELD
★★★½☆☆☆

EASY TO GROW
8/10

ZONE 7: SHADY AND DRY

Alpine strawberries
Fragaria vesca

	J F M A M J J A S O N D
SOW	▬▬▬
TRANSPLANT	▬▬▬
HARVEST	▬▬▬▬▬▬▬

FROM SOWING TO HARVEST
6–20 months

VARIETIES

'Golden Alexandra' – striking, yellow-green leaves, red fruits

'Mara des Bois' – traditional, tasty, larger red fruits

'Reine de Vallées' – Healthy, bushy, fewer runners, red fruits

'Pineapple Crush' – Unusual, green foliage, white-cream fruits

ORIGINS
Cooler temperate regions

ALSO GROWS IN
Zones 1, 3, 4, 5, and 6

HARDINESS
Fully hardy

LIFE CYCLE
Perennial

YIELD
★★/☆☆☆☆☆

EASY TO GROW
8/10

This strawberry may not produce the largest of fruits, but is still widely grown for its intense flavour, the berries prized by gourmet growers. Add to this their ability to thrive in tricky conditions, and it's hard to justify not growing them.

Growing essentials

Sowing Sow on the surface of moist compost, under cover in spring. Once large enough to handle, move seedlings into individual pots, and grow on until well established.

Planting Harden off and transplant outside, or buy potted plants in spring or autumn. Plant at 20cm (8in) spacings, into soil enriched with well-rotted organic matter.

Harvesting Attractive white flowers appear late spring and early summer, followed by tiny fruits. Harvest these individually when they part easily.

Secrets of success

Germination can take many weeks. Do not worry if seedlings fail to appear immediately.

Plants make excellent ground cover – a handy addition to weedy areas.

Consider growing other species *F. moschata*, *F. viridis*, and *F. chiloensis*.

Harvest regularly throughout summer, as fruits mature over many weeks.

Many forms produce runners – those that don't are best divided.

Fruits often hide amongst leaves so hunt thoroughly – pot growing gives easier access.

Alpine strawberries make excellent ground cover, thriving where conventional strawberries often struggle.

ZONE 7: SHADY AND DRY

Horseradish
Armoracia rusticana

	J	F	M	A	M	J	J	A	S	O	N	D
SOW			■	■	■							
TRANSPLANT						■						
HARVEST			■	■							■	■

TIME FROM SOWING TO HARVEST
32–40 weeks

VARIETIES
Available as the basic species

ORIGINS
Naturalized throughout Europe

ALSO GROWS IN
Zones 1, 2, 3, 4, 5, and 6

HARDINESS
Hardy

LIFE CYCLE
Perennial

YIELD
★★/☆☆☆☆

EASY TO GROW
9/10

Fans of Sunday roasts will no doubt already have a clump of horseradish on their plot. By joining them you will become the owner of a robust vegetable that needs very little care. Roots are renowned for their fiery flavour, and their vigorous nature allows plants to fend for themselves.

Growing essentials

Propagating Usually propagated via thick, dormant root divisions rather than seed, which takes a while to bulk up. Source from a healthy plant, or from specialist suppliers.

Planting Plant divisions in winter, into well-prepared soil, at 30 × 30cm (12 × 12in) spacings. Water well until established – mature clumps need little care. Young plants are also available at garden centres or herb nurseries. Plant in spring or autumn.

Harvesting Harvest roots when dormant, from late autumn to early spring. Prise up mature clumps and sever suitable roots, leaving enough intact for plants to re-grow. Roots can be 2–4cm (¾–1½in) thick on very established plants.

Secrets of success

Leaves are strong and robust, but inedible due to their pungency.

White flowers appear in summer – these are pleasantly scented and attractive to beneficial pollinators.

Horseradish is an incredibly versatile plant, with a strong, robust taproot that can extract moisture from deep within the soil.

You can sow seed, but emergence can be erratic and bulking up is slow.

Limit root spread by planting in large sunken containers.

Fresh young growth is very palatable to slugs and snails – take precautions to protect it.

Prepare horseradish sauce by grating fresh roots and using it as soon as possible. Do not cook as the pungency is lost.

Hazelnuts
Corylus species

	J F M A M J J A S O N D
PLANT	▬▬▬ ▬▬▬
HARVEST	▬▬

FROM PLANTING TO HARVEST
1–4 years

VARIETIES
C. maxima 'Ennis' – productive, large nuts, a good pollinator

C. m. 'Hall's Giant' – very hardy filbert, strong pollinator, good-sized nuts

C. avellana 'Butler' – vigorous, productive, flavoursome nuts

C. a. 'Red Majestic' – striking rich purple foliage, compact

ORIGINS
Naturalized in cooler regions of Europe, and tolerant of a wide range of conditions

ALSO GROWS IN
Zone 1, 3, 4, 5, and 6

HARDINESS
Fully hardy

LIFE CYCLE
Perennial

YIELD
★★/☆☆☆☆☆

EASY TO GROW
8/10

Whether you're a fan of cobnuts (*C. avellana*) or filberts (*C. maxima*), these husk-encased hazelnuts bring productivity and aesthetics to your garden. The trees are easily pruned to a manageable size, and the nuts offer a nutritious source of vitamins and protein.

Growing essentials

Planting Plant trees when dormant into weed-free soil improved with garden compost or well-rotted manure. Lay an organic mulch around roots to help trees settle in.

Planting Young trees need little care – water in very dry conditions to help the root system establish. Leave mature plants to their own devices, or prune to keep to a more modest size.

Harvesting Pick "green" nuts as soon as the kernels reach a good size in summer. Alternatively, leave nuts to turn brown on the tree, for longer storage until needed.

Filberts are incredibly low maintenance once trees are established.

Secrets of success

Hardiness varies – *C. maxima* is hardier so best for exposed sites.

Complete structural pruning in winter. Remove larger stems at the base to restrict tree height.

Boost pollination by growing more than one named variety, or by growing the species, too.

Green nuts taste juicy, sweet, and mild, and mature ones nuttier.

Squirrels can be thwarted by harvesting as soon as nuts are ready (dry in storage, if needed).

Use prunings on the plot – larger ones make excellent beanpoles and twigs can make woven supports.

ZONE 7: SHADY AND DRY

Japanese wineberry
Rubus phoenicolasius

	J	F	M	A	M	J	J	A	S	O	N	D
PLANT			▬	▬	▬						▬	▬
HARVEST							▬	▬				

FROM PLANTING TO HARVEST
2–3 years

VARIETIES
Available as the basic species

ORIGINS
Mountainous Asia. This vigorous scrambling plant can support a good crop, even in tougher conditions

ALSO GROWS IN
Zone 1, 3, 4, 5, and 6

HARDINESS
Fully hardy

LIFE CYCLE
Perennial

YIELD
★★★/☆☆☆☆

EASY TO GROW
8/10

This bramble has captured the gardening public's imagination, due to its highly ornamental value. The stems and unopened buds carry a thicket of bushy red bristles, and the berries ripen to a beautiful shimmering crimson. Add to that the rich, refreshing taste, and you're onto a winner.

Growing essentials

Planting Plant canes when dormant, into weed-free soil enriched with well-rotted organic matter. Mulch plants and water well to establish strong, resilient roots.

Training Train canes up and over garden buildings, or along fences and walls using a system of vine eyes and/or wires (see pages 74–75). Tie in while still young and flexible.

Harvesting Harvest fruit once fully coloured, gently rolling off the plant between finger and thumb. Pick over plants for 2–3 weeks as fruits ripen individually and gradually.

Secrets of success

Train into intricate shapes like spirals, hearts, and waves.

For the best flavour and juice, allow to ripen fully on the plant.

Propagate by tip layering – allow stem tips to root into the soil, then sever and replant.

Prune in autumn, removing fruited canes completely then tie new ones into place.

Mature plants grow to 6m (20ft) wide and 3m (10ft) tall – afford them plenty of space.

To raise from seed sow on the compost's surface, bag, and refrigerate for 2 months then place in a cold frame to germinate.

Japanese wineberries are highly ornamental plants that give trouble-free harvests.

Garlic mustard
Alliaria petiolata

	J F M A M J J A S O N D
SOW	
TRANSPLANT	
HARVEST	

FROM SOWING TO HARVEST
12–16 weeks

VARIETIES
Sold as the basic species

ORIGINS
Shady woodland margins in Europe and Asia

ALSO GROWS IN
Zone 1, 2, 5, and 6

HARDINESS
Hardy

LIFE CYCLE
Biennial

YIELD
★★★★½☆☆☆☆

EASY TO GROW
9/10

Drive along hedgerow-flanked country lanes during April and you'll see this plant growing in its natural habitat. A forager's highlight, garlic mustard, also known as Jack-by-the-hedge and hedge garlic, can also be cultivated in gardens and allotments to enjoy its flavoursome leaves during many other months of the year.

Growing essentials

Sowing If sowing undercover, to overcome seed dormancy: sow and keep at 20°C (68°F) for a fortnight, then bag and place in a fridge for 1 month, then return to 20°C (68°F) for germination. Transplant outside 15cm (6in) apart, once large enough to handle.

Seeds can be sown directly in spring or autumn. Create a seedbed, improving the soil with well-rotted organic matter. There is no need to thin resulting seedlings. Water well until established.

Harvesting Harvest leaves as and when required. Either pull up whole plants and defoliate, or pick individual leaves, taking care not to strip seedlings excessively.

Secrets of success

Spring sowings will give summer pickings, and autumn sowings harvest during early spring.

The flavour is a combination of mustard and garlic, mild when young, getting hotter with maturity.

Frothy white flowers will appear on mature plants – these are highly attractive to beneficial insects.

Foliage is an important food for caterpillars of orange-tip and green-veined white butterflies.

Eat foliage raw or cooked, as a salad plant or leafy vegetable. Flowers are also edible.

Direct-sown plants are most robust in dry soils, as they develop a strong taproot.

Being biennial, plants die after flowering, but then naturally self-seed in your garden.

Self-sown garlic mustard develops a drought-resistant root system.

Patience dock
Rumex patientia

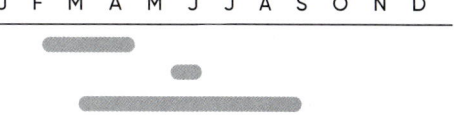

FROM SOWING TO HARVEST
20–22 weeks

VARIETIES
Available as the basic species

ORIGINS
European native, colonizing wasteland and field margins

ALSO GROWS IN
Zone 1, 3, 4, 5, and 6

HARDINESS
Hardy

LIFE CYCLE
Perennial

YIELD
★★★☆/☆☆☆☆

EASY TO GROW
9/10

The robust taproot of patience dock provides plants with excellent endurance in trickier sites.

Also known as monk's rhubarb and herb patience, it is an incredibly robust vegetable relied upon in leaner times for centuries. Revived through carbon farming methods, it provides harvests from February onwards in mild winters, due to its vigorous, reliable nature.

Growing essentials

Sowing If sowing under cover, sow onto the surface of seed compost and keep at 15–18°C (59–64°F). Keep compost moist but not waterlogged. Once large enough to handle, move individually into 9cm (3½in) pots, then plant out once bulked up at 30 × 30cm (12 × 12in) spacings.

Propagating Propagate via division. Take clumps from mature plants any time from October to March, replanting into well-prepared soil at 30 × 30cm (12 × 12in) spacings. Water young plants – mature ones need no extra care.

Harvesting Harvest leaves as and when required. Pull up whole stems and defoliate, or pick individual leaves. Flavour and texture intensify through the season.

Secrets of success

Leaves contain oxalic acid – don't eat too much if consuming raw. Cooking reduces levels significantly.

Mature plants are heavy yielding and incredibly useful for plugging the May "hungry gap".

Encourage soft and palatable foliage by shearing plants to soil level periodically throughout summer.

Plants will self-seed – remove flower spikes before they mature if you'd like to deter this.

Fresh growth is attractive to slugs and snails – take precautions to protect it during any wet weather.

Keep plants productive by lifting and dividing every 4–5 years, in winter.

ZONE 7: SHADY AND DRY

Perennial nettle
Urtica dioica

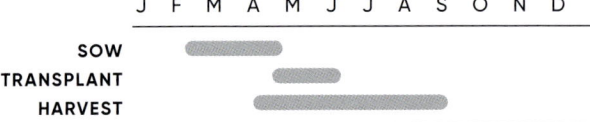

	J F M A M J J A S O N D
SOW	▬▬▬
TRANSPLANT	▬▬▬
HARVEST	▬▬▬▬▬▬▬

FROM SOWING TO HARVEST
16–20 weeks

VARIETIES
Sold as the basic species

ORIGINS
Temperate Europe and Asia, now found worldwide, due to its vigorous nature

ALSO GROWS IN
Zone 1, 3, 4, 5, and 6

HARDINESS
Hardy

LIFE CYCLE
Perennial

YIELD
★★★★½/☆☆☆☆

EASY TO GROW
10/10

Overlook the negative reputation of this plant and you'll be introducing yourself to a robust, vigorous, low-maintenance, and trouble-free edible that will crop on your plot for years to come. The leaves are nutrient-rich and can be managed to retain palatability. Nettles are also an important butterfly plant.

Growing essentials

Sowing Under cover, sow onto the surface of seed compost and keep at 15–18°C (59–64°F). Germination can be rapid, or erratic. Once large enough to handle, move individually into 9cm (3½in) pots, then plant out once bulked up at 30 × 30cm (12 × 12in) spacings.

Propagation Propagates readily via division. Take clumps from established plants any time October through March, replanting into well-prepared soil at 30 × 30cm (12 × 12in) spacings.

Harvesting Harvest as and when required. Either pull up and defoliate whole stems, or pick individual leaves. Their texture becomes more fibrous through the season.

Secrets of success

Other species are edible, but less palatable, like *U. urens*, *U. thunbergiana*, and *U. dioica* subsp. *gracilis*.

Perennial nettle is food plant for many caterpillars including peacock, tortoiseshell, red admiral, and comma.

This vigorous perennial edible will thrive in dry shade.

Eat the leaves as a vegetable or use them to make nettle beer and herbal teas.

Plants are either all male or all female – use this to prevent excessive self-seeding.

To stop leaves turning fibrous shear clumps back every 6 weeks to encourage fresh foliage.

Deter any invasive behaviour by growing clumps in large sunken containers, and removing the flower spikes.

Bamboo
Various species

ZONE 7: SHADY AND DRY

	J	F	M	A	M	J	J	A	S	O	N	D
PLANT				●					●			
HARVEST					▬▬▬▬							

FROM PLANTING TO HARVEST
1–3 years

VARIETIES
Phyllostachys edulis – spreading, sizeable species, 6m (20ft) tall; keep contained

Bambusa aurea – clump-forming, 3m (10ft) high green canes, yellowing with age

Sasa palmata – compact, spreading, 2m (6½ft) tall, broad green leaves

Chusquea culeou – clump-forming, 3m (10ft) tall, mid-green canes

ORIGINS
Throughout Asia, enjoying a wide range of conditions

ALSO GROWS IN
Zone 1, 3, 4, 5, and 6

HARDINESS
Half to fully hardy

LIFE CYCLE
Perennial

YIELD
★★/☆☆☆☆

EASY TO GROW
8/10

Bamboos are a diverse group, so one is surely suitable for your plot. These super-sized grasses are ornamental, coming in a range of colours, diameters, and heights. The tender new shoots are prized in Asian cuisine, and bamboo's trouble-free nature is perfect for low-maintenance gardens.

Growing essentials

Planting Plant while soil is warmer and moist, in spring or autumn. Bamboos are hungry so enrich the bed with well-rotted manure or garden compost, and general-purpose feed.

Maintaining Allow clumps to bulk up before harvesting. Young plants benefit from watering during dry spells, until established. Feed mature plants annually, in spring.

Harvesting Harvest from new canes as they push through the ground in spring and early summer. Leave a good proportion of new canes to mature.

Secrets of success

Bamboos are happy in most soils, except for waterlogged.

Mulch annually – their fallen leaves contain silica which helps to build resilient plants.

Grow in large pots or sink thick and deep barriers around plants, to control spread.

Clearing away the lower leaves can create a feature of the ornamental canes.

Divide plants if possible, every few years, in spring, to keep productive.

Boil rapidly before eating via rapid boiling, as the shoots contain toxins when raw.

Mature bamboos will produce an annual spring harvest, with plants establishing well in trickier sites.

Ten other star performers

Worcesterberry and jostaberry

Of all the *Ribes*, *R. divaricatum* and *R.* × *nidigrolaria* are excellent candidates for dry shade. Both are hardy and deciduous, best described as blends of gooseberry and blackcurrant. The marble-sized fruits are black once ripe.

Blackberries and hybrid berries

A large, diverse, and vigorous group of scrambling cane fruits, many are supremely hardy, with generous yields. Blackberries (*Rubus fruticosus*) and hybrids (predominantly the *R.* Tayberry Group and *R.* × *loganobaccus*) excel in dry shade. Compact forms are becoming increasingly available.

Honeyberry

Very hardy deciduous climber, *Lonicera caerulea* is found throughout the Northern Hemisphere. Best grown in twos or threes for good yields, the elongated berries are dusky blue with a tangy sweet-sour flavour. Scented flowers in spring.

Sweet cicely

A perennial herb with the most beautiful aniseed scent, sweet cicely (*Myrrhis odorata*) is most drought resistant when allowed to self-sow. Leaves are produced year-round, and frothy flowerheads, loved by insects, appear in summer. The roots and seeds are also edible.

Wild garlic

A European native, *Allium ursinum* or ramsoms is frequently found in moist woodlands. A rather over-familiar plant in damp shade, drier areas will limit spread via bulbils and seed. Very hardy, producing attractive white flowers in early summer.

The long thorny stems of blackberries scramble up to the light when grown through trees and shrubs.

Large swathes of wild garlic can develop on damp soils, but in drier beds vigour is more controlled.

Land cress

A hardy European biennial or short-lived perennial *Barbarea verna* offers beautifully peppery leaves year-round. Bulks up quickly from seed to form an attractive rosette of scalloped leaves. Self-sown, mulched plants offer most drought resilience.

Udo

An East Asian hardy perennial (*Aralia cordata*) with striking ornamental leaves. Blanch young shoots in spring as a mild-tasting "asparagus". Mulch young plants to bulk up and build drought resistance. 'Sun King' bears golden foliage.

Barberries

Various *Berberis* species give us barberries – *B. vulgaris*, *B. wilsoniae*, *B. aristata*, *B. buxifolia*, and *B. darwinii*. Tart fruits (cook with sugar) grow abundantly on thorny shrubs, some deciduous, others evergreen. Springtime flowers are excellent for pollinators.

Oregon grape

Many forms of *Mahonia* – *M. × media*, *M. aquifolium*, *M. nervosa*, and *M. repens* – offer Oregon grapes. The dusky purple berries taste sweet once ripe, but can also be cooked. Spiky evergreen foliage, with springtime yellow flower clusters are invaluable for bees.

Plum yews

Two species of *Cephalotaxus*, *C. harringtonia*, and *C. fortunei*, produce tasty cherry-sized fruits. These East Asian hardy conifers are slow-growing and all-male or all-female – "dioecious" – source one male selection to pollinate multiple females.

The leaves of land cress are a welcome addition to spring and autumn salads.

Barberries are trouble-free shrubs offering attractive springtime flowers and autumn fruits.

Project:
Making a miracle mulch for healthy roots

Deep shade usually marries with damp soil, but occasionally, alongside a large hedge or tree, dry conditions predominate. Many plants happily grow here, but to support them further, why not create a moisture-retaining mulch from compostable items? It's free – and it's a cinch.

Inheriting a sizeable tree or hedge in your new home can be incredibly useful. Its presence brings structure and offers a backbone to younger plantings. Alongside this, you also inherit a large root system and if this shares soil with your fruit and vegetable beds, this creates competition for water. The mature plant will invariably hold the upper hand due to its established root network.

Mimicking nature

All plants listed in Zone 7 will hold their own in shady, dry conditions, but if you'd like to support them further or grow edibles that prefer moist shade, create a mulch. Organic mulches mimic nature: in woodlands where root competition is fierce, a natural mulch is created via falling foliage, annually in autumn in deciduous woodlands, or gradually over the year in coniferous forests, as needles fall. We often remove leaf litter when we tidy our gardens, adding a mulch puts it back. The result is a network of insulated, moistened, and sustained healthy roots.

ZONE 7: **SHADY AND DRY**

You will need

Garden and kitchen waste
Compost area
Garden shredder or shears
Garden fork

Steps

1 Your compost area can simply be a pile on a spare patch of ground, or a system of bins. Adopt the mindset of recycling all garden waste. Ideally, pass the material through a garden shredder before composting, or chop woodier pieces into smaller fragments.

2 Add garden waste to your compost heap whenever you generate it, aiming to add a mixture green (sappy) and brown (woody) items together rather than in separate layers.

3 Regularly turn your heap to mix its contents (at least four times per year) for even decomposition as all components spend time in the middle (the hottest, most efficient part).

4 Once the compost is dark brown and no plant parts are distinguishable (generally after 9–12 months) add the mulch to your edible beds. Lay it at least 5cm (2in) thick so that crops feel the benefits, ideally in spring.

Not having access to an outdoor space shouldn't limit your grow-your-own aspirations – by cultivating edibles undercover you have total control over their growing environment. No hurricanes, no floods, no frosts – how fantastic is that? Say hello to space-saving saucers of pea shoots, wheatgrass, salad leaves, and microgreens. Home-grown lemongrass, stevia, ginger, turmeric, galangal, and cardamom can grace your kitchen chopping board, while lemons, limes, and kumquats will get heads buzzing with dessert and drink recipe ideas. Compact chillies and tomatoes will enliven your cooking repertoire, and how could you not feel decadent supping on a cup of home-grown green tea?

Zone 8
Indoors

ZONE 8: INDOORS

Dwarf tomatoes

Solanum lycopersicum

	J	F	M	A	M	J	J	A	S	O	N	D
SOW		▬	▬	▬								
TRANSPLANT			▬	▬	▬							
HARVEST						▬	▬	▬	▬	▬	▬	▬

FROM SOWING TO HARVEST
14–20 weeks

VARIETIES

'Veranda Red' F1 – stocky, upright plants, productive, red cherry-sized tasty fruits

'Losetto' F1 – cascading, red cherry tomato, ideal for hanging baskets, blight resistant

'Tiny Temptations Yellow' F1 – cascading basket type, yellow cherry-sized fruits

'Lemon Sherbert' F1 – upright, bushy plants yielding yellow, cherry-sized fruits

ORIGINS
Mexico and the Americas

ALSO GROWS IN
Zone 1, Zone 2, and Zone 3 – if sufficiently warm

HARDINESS
Not hardy – killed by frosts

LIFE CYCLE
Perennial, grown as an annual

YIELD
★★★★/☆☆☆☆

EASY TO GROW
9/10

In its original form the tomato is a lax, sizeable, sprawling plant. Selection has allowed more compact forms to appear, and these days there is a concerted effort to breed more fruit colours, shapes, and disease resistance into compact tomato varieties. The result is a fantastic range of tiny tomatoes.

Growing essentials

Sowing Sow seeds 5mm (¼in) deep, 10mm (½in) apart, in pots under cover in spring. Once seedlings are large enough to handle, transplant individually into 9cm (3½in) diameter pots and grow on under cover.

Planting When the first tiny flower trusses appear, pot on into the final container. Its size should accommodate the final height and spread of your plants – if in doubt, choose a larger pot.

Harvesting Tomatoes enjoy bathing in sunshine. Water well, especially as roots mature, and feed regularly with a high-potash, organic liquid tomato fertilizer. Harvest once ripe.

Secrets of success

Set propagators to 18–22°C (64–71°F) for good germination. Expect seedlings in 7–10 days.

For indoor cropping sow as early as February – ensuring seedlings receive enough light.

Compact tomatoes such as 'Losetto' can yield prolifically from a small space.

Growing lights are especially useful for boosting growth early and late in the season.

Buy plants if you can't sow – the range available is ever-increasing.

Compact tomatoes often yield well initially, growth then tails off, with later harvests occurring in flushes.

Consider sowing in batches for year-round cropping indoors.

Pea shoots
Pisum sativum

	J	F	M	A	M	J	J	A	S	O	N	D
SOW	●	●	●	●	●	●	●	●	●	●	●	●
HARVEST	●	●	●	●	●	●	●	●	●	●	●	●

FROM SOWING TO HARVEST
2–4 weeks

VARIETIES
Any pea variety can be grown as pea shoots

ORIGINS
By growing peas for their fresh shoots, indoors, you encourage the sweetest, most tender growth

ALSO GROWS IN
Zone 1 and Zone 3

HARDINESS
Hardy

LIFE CYCLE
Annual

YIELD
★★★★/☆☆☆☆

EASY TO GROW
9/10

If you want a deliciously fresh, family-favourite harvest in a matter of weeks, then look no further than pea shoots. These vigorous little plants carry exactly the same flavour as their podded counterparts, yet they're ready to cut and eat in a fraction of the time. Pea shoots are incredibly easy to grow and a windowsill go-to.

Growing essentials

Sowing Fill a suitable container with a 1–2cm (½–¾in) layer of compost – you can use multi-purpose, rather than more costly seed compost. Water well, allow to drain, then sow seeds thickly on top.

Maintaining Keep compost moist while seeds are germinating and shoots lengthen. Do not overwater – tip any surplus liquid away. Turn trays often to avoid legginess.

Harvesting Cut with scissors, once shoots are 8–12cm (3–4¾in) tall. Eat promptly for the sweetest flavours – best refrigerated for only a few hours, or kept in iced water.

Secrets of success

Dried marrowfat peas from supermarkets offer an inexpensive source of seeds.

Any shallow container is suitable – experiment with recycling receptacles.

Compost isn't essential – cotton wool or kitchen paper can also support pea shoots.

Pea shoots stand well but cut them before they get too old otherwise starch levels build.

Sow a succession of containers every week or so, for continual harvests.

Sowings resprout after their initial harvest, giving a second, smaller crop.

A small saucer of pea shoots will take up little room, yet offer good yields.

Microleaves
Various species

	J F M A M J J A S O N D
SOW	▬▬▬▬▬▬▬▬▬▬▬▬
HARVEST	▬▬▬▬▬▬▬▬▬▬▬▬

FROM SOWING TO HARVEST
1–4 weeks

VARIETIES
Coriander 'Micro Coriander Splits' – quick-to-germinate form of this popular leaf

Basil 'Lemon' – a pairing of lemon and basil flavours

Chervil – germinates quickly, leaves have a delicate aniseed flavour

Agastache foeniculum – rapid and even emergence of seedlings, warm, mint-like taste

ORIGINS
Various

ALSO GROWS IN
Best grown in pots or trays

HARDINESS
Various

LIFE CYCLE
Various

YIELD
★★/☆☆☆☆☆

EASY TO GROW
9/10

These leaves marry miniature harvests with maximum flavour. Their compact and speedy nature makes them popular with all gardeners, no matter the size of plot. They don't actually need soil at all! A windowsill can become the ultimate flavour generator for your kitchen.

Growing essentials

Sowing Fill a suitable container with a 1–2cm (½–¾in) layer of compost. Water well, drain, then sow your seeds on top. Cover with a fine layer of compost and water gently.

Planting Position in good light – artificial light can help (see pages 186–187). Keep compost just moist during germination. Do not overwater, tip surplus liquid away.

Harvesting Harvest, using scissors, once shoots are 1–3cm (½–1¼in) tall. Eat as soon as possible to enjoy the most intense flavour – wash and store leaves in the fridge for a few days, if needed.

Secrets of success

Try also rocket, fennel, mustard, mint, chives, perilla, sorrel, radish, oxalis, beetroot, and amaranth.

Any shallow container is suitable, so experiment with recycling various receptacles.

Compost isn't essential – cotton wool or kitchen paper also supports microleaves.

Hand misting is an easier alternative to watering.

Sow a succession of containers every week or so, for continual harvests.

Flavour changes with maturity, taste-test before harvesting.

Microleaves need little space but they can bring strong flavours.

Prickly pear
Opuntia species

ZONE 8: INDOORS

	J F M A M J J A S O N D
PLANT	▬▬▬▬▬▬▬▬▬▬▬▬
HARVEST	▬▬▬▬▬▬▬▬▬▬▬▬

FROM PLANTING TO HARVEST
1 week–several years

VARIETIES
O. bergeriana – large leaves, small red fruits, 1.5m (5ft) tall

O. ficus-indica – large leaves, orange-red fruits, 2m (6½ ft) tall

O. polyacantha – compact, very spiny green paddles, yellow or cerise blooms

O. robusta – almost spherical leaves, occasional small red fruits, to 1m (3ft)

ORIGINS
The deserts of America, now found throughout arid regions

ALSO GROWS IN
Zone 1 and Zone 2, for summer

HARDINESS
Tender

LIFE CYCLE
Perennial

YIELD
★★/☆☆☆☆

EASY TO GROW
9/10

Recognizable as the master of desert life, you can grow these statuesque plants in your home. The paddles and fruits are incredibly flavoursome for both sweet and savoury dishes. Add to this their ultimate, low-maintenance needs, and you're onto a winner.

Growing essentials

Planting *Opuntia* require excellent drainage, pot into a 50:50 mixture of peat-free compost and grit or sharp sand. Larger pots produce taller plants. Position in full sun.

Aftercare Plants are drought tolerant – avoid excess watering. Add high potash liquid feed every fortnight during growth, for flowers and fruits.

Harvesting Harvest paddles as soon as large enough – younger ones have softer spines. Flowers appear in summer, and fruits ripen in autumn – harvest just as they soften.

Secrets of success

Pot-grow for compact plants (many would grow taller otherwise).

To remove spines, use a sharp knife, grill lightly, or rub with thornproof gloves.

More compact species are edible, like *O. microdasys*, but have dense spines.

For fruit production, mature *O. ficus-indica* is the most reliable.

Opuntia can make large specimens in their native habitat, but we can grow them as more compact houseplants in full sun.

Propagate via cuttings – remove paddles, leave to callus, then anchor into gritty compost.

Sow in spring into gritty compost, in a well-lit propagator set at 18–20°C (64–68°F).

Lemongrass

Cymbopogon species

	J F M A M J J A S O N D
SOW	▬▬▬
TRANSPLANT	▬
HARVEST	▬▬▬▬▬▬▬▬▬▬▬

FROM SOWING TO HARVEST
22–32 weeks

VARIETIES
C. citratus – widely available, 1m (3ft) tall clumps

C. flexuosus – more compact, better suited to smaller containers

C. nardus – grown for citronella oil, also a useful flavouring

ORIGINS
Widely distributed throughout many tropical regions

ALSO GROWS IN
Zone 1, for summer

HARDINESS
Tender

LIFE CYCLE
Perennial

YIELD
★★★★/☆☆☆☆

EASY TO GROW
8/10

A backbone of Asian cuisine, these sizeable and highly aromatic grasses easily supply you with all the stems needed for self-sufficiency. Their rapid growth and striking aesthetics bring ornamental value to your home, and their trouble-free nature makes them a welcome, low-maintenance guest.

Growing essentials

Sowing Sow seeds on the surface of a pot of moist compost, placing it in a heated propagator set to 18–22°C (64–71°F) until well emerged. Once large enough to handle, transplant individually into pots.

Planting Seedlings bulk up quickly, move into larger pots of multi-purpose compost as the roots develop. Water well and liquid feed with a balanced formula every week.

Harvesting Harvest stems as soon as their bases become pencil thick. Sever from the plant, ensuring plenty of stems remain, trimming off brown outer layers.

Secrets of success

Divide clumps when congested using an old breadknife, to ensure plants remain productive.

Plants can live outside in summer but must be somewhere frost-free for winter.

Seed will germinate readily – ample warmth and moisture is key.

Lemongrass is wind tolerant so great for a balcony garden.

Just one lemongrass plant will provide sufficient harvests for a household.

Take cuttings, ideally with some roots intact. Shop-bought stems also root successfully.

Foliage is tough – wear stout gloves when combing plants to remove dead leaves.

Citrus

Citrus species

ZONE 8: INDOORS

	J	F	M	A	M	J	J	A	S	O	N	D
PLANT			▬	▬	▬	▬	▬	▬	▬	▬	▬	
HARVEST	▬	▬	▬							▬	▬	▬

FROM PLANTING TO HARVEST
1 week–several years

VARIETIES

C. australasica – caviar lime, elongated fruits, relatively hardy trees

C. hystrix – wrinkled limes, highly aromatic leaves

C. japonica – kumquat, strong-growing, productive

C. × limon – vigorous, healthy hybrid, good-sized lemon trees

ORIGINS
Tropical regions, plants enjoy well-lit, frost-free locations

ALSO GROWS IN
Zone 1, for summer

HARDINESS
Half hardy to frost tender

LIFE CYCLE
Perennial

YIELD
★★★/☆☆☆☆☆

EASY TO GROW
5/10

C. reticulata produces numerous small, highly aromatic and flavoursome satsumas.

Reminiscent of Mediterranean holidays and thirst-quenching drinks, citrus fruits are packed with zest, juice, flavour – and that unmistakable aroma. In their native climes they develop into sizeable trees. In pots indoors we tame them into compact yet productive plants.

Growing essentials

Planting Pot into compost that holds nutrients yet drains freely. Use a citrus compost, or blend peat-free and John Innes ericaceous composts with grit.

Aftercare Water trees regularly and provide a buoyant, humid atmosphere when in full growth. Feed regularly with an appropriate fertilizer. Watch for red spider mite and scale insect.

Harvesting Highly scented, cream-coloured flowers appear in spring, followed by fruits that ripen during autumn and winter. Harvest once well coloured and full sized.

Secrets of success

Citrus winter food is balanced in nutrients, whereas summer food is high in nitrogen.

Overwintering outdoors is only possible for hardier citrus like *C. trifoliata* in sheltered gardens.

Fruits will store successfully for many weeks if refrigerated.

Pruning consists of thinning out congestion in winter, and pinching out new shoots in summer.

Citrus can be raised from seed but quality is variable and fruiting takes many years.

Trees enjoy a spell outside during summer – be sure to acclimatize them to full sun.

Ginger and relatives

Zingiber, Curcuma, Elettaria, and *Alpinia*

	J F M A M J J A S O N D
PLANT	▬▬▬▬▬▬▬▬▬▬▬▬
HARVEST	▬▬▬▬▬▬▬▬▬▬▬▬

FROM PLANTING TO HARVEST
1–several weeks

VARIETIES

Zingiber officinale – creeping rhizomes, fiery ginger flavour, broad green leaves

Curcuma longa – rich yellow turmeric rhizomes, mid-green leaves

Alpinia galanga – vigorous pale cream galangal rhizomes, robust rich green foliage

Elettaria cardamomum – rhizomatous perennial, cardamom seed pods

ORIGINS
India. Plants have a healthy appetite for food, warmth, and water

ALSO GROWS IN
Zone 1, during summer

HARDINESS
Tender

LIFE CYCLE
Perennial

YIELD
★★★/☆☆☆☆

EASY TO GROW
8/10

The ginger family contains a range of important spices – it's exciting that you can grow them at home. Whether you choose to cultivate the flavoursome rhizomes of ginger, turmeric, or galangal, or the aromatic seed pods of cardamom, all will benefit your kitchen.

Growing essentials

Planting Plant rhizomes shallowly in large pots – 25cm (10in) diameter minimum – of peat-free multi-purpose compost enriched with general-purpose fertilizer. Water gently.

Aftercare As growth advances, increase watering and feed each week with balanced liquid fertilizer. Plants appreciate being positioned in shade or semi-shade.

Harvesting Harvest rhizomes of ginger, galangal, and turmeric as soon as they are large enough – leave enough in place for growth to continue. Harvest cardamom seedpods once fully hardened.

Secrets of success

All are hungry feeders – feed weekly during active growth.

Plants are happier in shade than many other edible houseplants, making them useful gap fillers.

Repot or divide plants if overgrown in summer.

The flowers are attractive so be sure to appreciate them when they appear.

Ginger makes a striking foliage plant that grows happily in shade.

Plants become dormant in winter – remove faded stems, and reduce watering.

Propagate by potting up healthy sections of shop-bought rhizomes.

ZONE 8: INDOORS

Pomegranate
Punica granatum

	J F M A M J J A S O N D
PLANT	▬▬▬▬▬▬▬▬▬▬▬▬
HARVEST	▬

FROM PLANTING TO HARVEST
1 week–several years

Popular as a health-boosting fruit that is packed with antioxidants, pomegranates are perfect for a sunny room or conservatory. Their slow rate of growth makes these trees easy to accommodate, and the showy flowers and fruits are highly ornamental.

Growing essentials

Planting Pomegranates enjoy a free-draining yet fertile compost – pot into a mixture of loam-based compost, peat-free compost, and grit. Water gently in spring.

Aftercare Water trees freely when in active summer growth – feed fortnightly with a high potash liquid fertilizer. Fierce sun or wind can cause leaf scorch.

Harvesting Flowers appear on mature plants in early summer, followed by spherical fruits. These appreciate ample warmth in autumn to ripen. Harvest once fully formed – fruits store for a week or so.

Secrets of success

Trees can overwinter outside in sheltered plots – insulate pots and shoots against frosts.

For maximum fruiting feed regularly, avoid over-pruning, and give ample autumn warmth.

Repot root-bound trees in spring, ensuring pots have ample drainage holes.

Sow 1cm (½in) deep in moist seed compost, at 18–24°C (64–75°F) till emerged.

Keep pruning to a minimum – remove any dieback or congestion in spring.

Only some flowers develop into fruits. Trees are self-fertile but cross-pollination helps fruit set.

VARIETIES
P. granatum – widely available, bushy habit, red-flushed fruits

P. g. **var.** *nana* – compact, ideal for smaller containers

P. g. 'Mollar de Elche' – seedless, orange-skinned fruits, juicy red flesh

P. g. 'Wonderful' – healthy plants, good-sized red fruits

ORIGINS
Southern Asia. Now widely grown in tropical climates with long, sunny summers

ALSO GROWS IN
Zone 1, for summer

HARDINESS
Half hardy

LIFE CYCLE
Perennial

YIELD
★★½/☆☆☆☆

EASY TO GROW
6/10

It is possible to grow good-sized pomegranates as houseplants.

Agave
Various species

	J F M A M J J A S O N D
PLANT	▬▬▬▬▬▬▬▬▬▬▬▬
HARVEST	▬▬▬▬▬▬▬▬▬▬▬▬

FROM SOWING TO HARVEST
1 week–several years

VARIETIES
A. americana – sizeable open rosette of strap-like spiny leaves

A. parryi – dense rosette, spoon-shaped glaucous, spine-tipped leaves

A. salmiana – very large, urn-shaped agave, broad grey-green leaves

A. utahensis – compact dome of narrow glaucous, spine-tipped leathery leaves

ORIGINS
Hot, arid regions of the Americas

ALSO GROWS IN
Zone 1 and Zone 2, for summer

HARDINESS
Tender to half hardy

LIFE CYCLE
Perennial

YIELD
★★/☆☆☆☆

EASY TO GROW
9/10

Hugely popular due to their good looks and easygoing nature, agaves also offer value in the kitchen, as plant hearts, flowers, stems, and sap are all edible. So get ready to roast plants and condense sap, as you embrace all that agaves offer.

Growing essentials

Planting Use a free-draining mix of 50:50 peat-free compost and sharp sand for good drainage. Position in full sunlight.

Maintaining Agaves are naturally drought tolerant so avoid excess watering. Add a liquid feed every fortnight during active growth. Remove any brown and withered leaves.

Harvesting Harvest from mature plants. Removing hearts for roasting and sap collection are quite destructive processes – ensure you have spare young plants available for future cropping.

Secrets of success

Many agaves are edible (*A. vivipara*, *A. tequilana*) but some find the sap irritating.

Plants can live outdoors for the summer, but bring them in for winter.

Agaves grow well from seed, an inexpensive way to build a collection.

Many agaves bear vicious spines – position out of reach of children.

Large flower spikes can appear on mature plants – this particular rosette will then die.

Agave "pups" – small rosettes at the base of plants – can be severed and used for propagation.

Agaves appreciate living in a sunny spot outdoors for the summer months.

ZONE 8: INDOORS

Dwarf chilli peppers
Capsicum species

	J	F	M	A	M	J	J	A	S	O	N	D
SOW			▬	▬								
TRANSPLANT				▬	▬							
HARVEST							▬	▬	▬	▬	▬	

FROM SOWING TO HARVEST
24–32 weeks

VARIETIES
'Apache' F1 – bullet-shaped, scarlet fruits, hot, productive

'Numex Twilight' – hot, teardrop-shaped, purple, red, orange, and yellow fruits

'Pixie Lights' – masses of upright, long fruits, purple ripening to red, very hot

'Dawn' – gently cascading plants, long, yellow fruits ripening to red, hot

ORIGINS
Sunny areas with minimal rainfall in central and southern America

ALSO GROWS IN
Zone 1, plus Zone 2 and Zone 3 if sufficiently warm

HARDINESS
Not hardy – killed by frosts

LIFE CYCLE
Perennial, predominantly grown as an annual

YIELD
★★★★★/☆☆☆☆☆

EASY TO GROW
9/10

The chilli's popularity as a potted plant has fuelled breeding of compact varieties and indoor gardeners benefit hugely from this. Some, like 'Stumpy', grow to only 20cm (8in) tall yet can be smothered in fruits. Others like 'Cardiff Queen' have a beautifully cascading habit on free-fruiting plants.

Growing essentials

Sowing Sow seeds 1cm (½in) deep, 1cm (½in) apart, in pots under cover in spring. Once seedlings are large enough to handle, transplant individually into 9cm (3½in) pots and grow on under cover.

Planting When the roots fill the pot, move plants into their final container – 1 litre (¼ gal) minimum. Larger pots give larger plants. Water well and feed regularly with high-potash liquid fertilizer.

Harvesting Harvest while green, or allow to colour up for sweeter and hotter flavours. Dry once mature to store for years.

Secrets of success

Set propagators to 18–25°C (64–77°F) for good germination – especially for the hottest chillies.

Gently shake indoor chilli peppers when in flower to aid pollination.

Choose final containers according to your varieties – some are incredibly compact.

You can treat as perennials indoors – plants will often crop year-round.

To overwinter, reduce watering and prune slightly. Re-pot in spring.

Preserve gluts by freezing mature and immature fruits in sealed bags.

Compact varieties of chilli such as this 'Numex Twilight' make very attractive and productive houseplants.

Ten other star performers

Mushroom plant
If you love mushroom flavour but not the texture, *Rungia klossii* is for you. This tender Indonesian evergreen's shiny green leaves have a crisp texture and striking mushroom taste. Happy in filtered sun or part shade.

Tea
Produce your own green or black tea with young leaves of this borderline hardy evergreen shrub. East Asian *Camellia sinensis* makes an attractive houseplant for shadier corners. This plant prefers ericaceous compost, and ample moisture, but not waterlogging.

False shamrock
Oxalis triangularis is a borderline hardy bulbous perennial, producing attractive purple, shamrock-style leaves. Pleasantly tart flavour. Also produces pretty white flowers and fleshy roots, both edible. Easily propagated via bulbils.

Wheatgrass
Raw food dieters can grow swathes of wheatgrass on their windowsills. Sow *Triticum aestivum* seeds onto shallow trays of damp kitchen paper every 2 or 3 weeks, harvesting, and juicing for a nutritious hit, once 10–12cm (4–4¾in) tall.

Barbados aloe
A widely distributed tropical plant easily grown in a pot of free-draining compost. *Aloe vera*'s fleshy leaves, composed of a bitter-tasting skin that must be peeled away to expose a gel-like pith. Blitzed with citrus juice and sugar, this pith makes a refreshing drink.

Candyleaf

Use the leaves of *Stevia*, known botanically as *S. rebaudiana*, a tender South American perennial, as a natural sweetener. It is readily raised from seed, and therefore treated as an annual. Pinch out to encourage bushy growth.

Scented pelargoniums

Use this tender perennial as a flavouring – look for *P. crispum*, *P. graveolens*, *P. odoratissimum* and *P. tomentosum*. Many named varieties are available too. Grow in pots of moist yet free-draining compost. Cuttings root readily.

Roselle hibiscus

Steep the dried calyx of this tender plant, *Hibiscus sabdariffa*, or Jamaican sorrel, to make a pleasant tea. The fresh flowers and leaves are also edible. Hailing from tropical Africa, it can be raised from seed and enjoys dappled shade.

Cut-and-come-again salads

In good light, various salad leaves – rocket, lettuce, chicory, endive, mustard, mizuna, pak choi, chard, spinach – grow to 12–15cm (4¾–6in) tall. Sow in shallow trays of seed compost. Harvest above growing points repeatedly as they re-sprout.

Lemon verbena

A borderline hardy shrub (*Aloysia citrodora*) bearing lemon-scented leaves used to flavour drinks and desserts. Pinch out to keep compact, or allow to develop panicles of small white flowers. Move outdoors for summer, if desired.

(Far left) Barbados aloe's fleshy leaves make it drought tolerant; (left) lemon verbena can grow in a sunny spot outside, for summer; (top) candyleaf makes an excellent, seed-raised houseplant; (above) *Pelargonium citronellum* enjoys a sunny, well-lit position.

Project: Creating an artificial light growhouse

Having no outdoor space need not stop you growing your own – simply move your cropping space indoors. Artificial lighting has progressed hugely in response to the demand for urban edible cultivation. The result is vertical growing – the ultimate, space-saving system for city dwellers.

This project is low footprint, high output, transforming a standard shelf unit into a stack of indoor growing zones. The unit needs to be durable and easy to clean. Secure short LED tubes (see pages 192–193) to the shelves, providing edibles with the appropriate light for healthy, stocky growth.

Indoor edibles

This growhouse makes an excellent propagation environment, allowing you to sow earlier in the year, when outdoor light levels are still low. Growing a range of crops to maturity is also possible. Look for those with a squat rosette of foliage. Many salad leaves are ideal for compact cropping; lettuce, rocket, land cress, purslane, mibuna, pak choi, and mustard. Leafy herbs, too, like basil, oregano, mint, coriander, and parsley will thrive, and couldn't be more conveniently placed for the kitchen. Even root crops such as 'Paris Market' carrots, radishes, turnips, and beetroot can grow indoors. Look for compact varieties such as pea 'Half Pint', chilli 'Basket of Fire', and tomato 'Veranda Red'.

ZONE 8: INDOORS

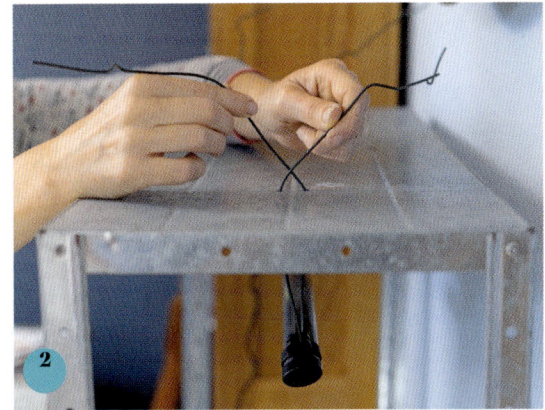
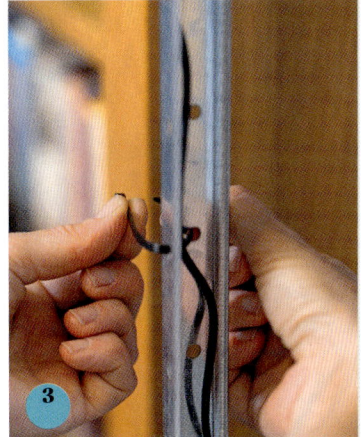

You will need

Shelving unit
Spirit level
Artificial light tubes
Wire or plastic clips
Trays

Steps

1 Position the shelving unit on hard (waterproof) flooring, ensuring it is level, so that when you add water the trays don't overflow. It is also useful to stand the whole system in a large waterproof tray.

2 Adjust the shelves so that they sit at least 30cm (12in) apart in height – closer spacings are possible (for propagation, for example) but this limits the range of crops that you can grow to maturity. Secure fluorescent tubes to the underside of each shelf using wire or plastic clips.

3 Secure electric cables to the sides of the shelving unit.

4 Position appropriately sized trays on each shelf. There is no need to clad the growhouse.

5 Cultivate your chosen edibles in sufficiently sized pots or trays, watering and feeding as required.

Tweaks to optimize your plot

Water is a key element in any garden, so managing it wisely will help you run your plot with supreme efficiency. Harnessing all that falls from the skies, delivering it to your crops in appropriate ways, and manipulating your soil so that it can store it effectively, will all help provide optimum harvests.

Water

Collecting and applying water with maximum efficiency

(Left) Capture as much rainfall as possible from the roofs of garden buildings on your plot. (Right) Self-watering systems are ideal for gardeners planning summer vacations away from the plot. (Below right) "Dipping tank"-style containers can make good use of non-cropped areas.

Rainfall capture

With many water meters compulsory in drought-stressed areas, it's logical to harvest rain. The average house roof captures an astonishing 50,000 litres (13,208 gallons) of water per year – so aim to capture at least some of this for your plot. A standard glasshouse or shed can collect 5,000 litres (1,320 gallons) per year; attach gutters, downpipes, and storage vessels wherever possible.

Place smaller 200-litre (52 gallon) water butts or tanks in areas that are tricky to access; I have two second-hand galvanized cattle troughs in a narrow passage behind my greenhouse. New, underground storage tanks are costly, but smaller 1,000-litre (264 gallon) reconditioned models (IBCs) are more affordable. Buried, they steal no growing space from your plot.

Irrigation

Propagation is a water-demanding period where seedlings and transplants require nurturing. If growing crops to maturity under glass, consider self-watering systems with integrated water reservoirs. They offer minimal water wastage and maximum yields. Drip irrigation is invaluable for container-heavy plots – ensure drippers have adjustable flow rates, to meet individual plants' needs. Bury porous soaker hoses 5–8cm (2–3¼in) in the soil, then mulch, to deliver water directly to the roots of fruit or vegetables in rows. Overhead sprinklers waste water and by wetting the soil surface encourage weeds (see pages 106–107 for weed control hints). Hosepipes are convenient – retractable models more so. Position stout poles on plot corners to guide them. If using grey water (domestic wastewater) on the plot, target it towards the soil surrounding crops that are cooked before consumption (not eaten raw), and avoid storing for more than 24 hours.

Soil management

Bare soil is rare in nature; mulching locks moisture alongside plant roots, where it's most needed. Work with your soil type (see pages 52–57). Adding organic matter (e.g. garden compost or well-rotted animal manure) boosts drainage on heavy soils and moisture retention on drier ones. Trenches wick away water after floods and make useful irrigation gullies during drought.

TWEAKS TO OPTIMIZE YOUR PLOT

All plants need ample light, especially during propagation. Greenhouse or polytunnel multi-directional light is ideal, but if you don't have either, consider artificial lighting. Harnessing sunlight in the garden is also becoming increasingly feasible via solar technology, and can significantly reduce your energy bills.

Light

Utilizing solar power and maximizing artificial sources

Indoor propagation

Many of us in flats have to deal with one-directional light when propagating our edibles. Light from one side, however strong, causes long, stretched growth, or "legginess" of seedlings, making them weak and vulnerable to collapse. Thankfully this is preventable. When you sow seeds or pot cuttings on your windowsill, reflect the light back by securing a simple low wall of cardboard or plywood behind your propagating space, painting the structure white or covering it with aluminium foil. Turning containers around every few days will also help.

Artificial light

If leggy seedlings are a problem, or you'd like to grow leafy crops indoors or under cover through the dark days of winter, investing in an additional light source can noticeably improve plant growth. Units are increasingly affordable and versatile. Plants respond favourably to specific spectrums of light: red for flowering growth, blue for leaf growth. LED lights are available in a range of spectrums from glasshouse equipment suppliers. They are energy efficient, long-lived, and give off little heat, compared to older lighting systems. Suspend lights from greenhouse staging, or create space-efficient growing houses from shelving (see pages 186–187) – ideal for urban dwellings.

Solar power

Consider using sunlight to power your propagation environment. Garage and shed roofs are underused spaces, and solar panels are becoming increasingly compact and efficient. In their simplest form solar panels come integrated into watering, heat, or light units. At present, model options are limited; look at off-grid or camping systems for general energy generation instead. You can then direct this to your lighting, heating, or watering systems.

TWEAKS TO OPTIMIZE YOUR PLOT 193

(Above left) LED lights offer an effective way to boost winter light levels.
(Above) Deter leggy windowsill seedlings by using a simple reflector of foil.
(Left) Compact solar panel systems utilize the sun's power.

While it's neither feasible nor desirable to adjust outside temperatures, your indoor propagation environment can benefit from added warmth in spring. One size doesn't necessarily fit all, so look at the various heat source options available to you. Those embracing global crops might create a frost-free winter environment too.

Temperature

Using propagators, heaters, and heat mats

(Above) A greenhouse offers the ideal propagation environment for your edibles.
(Above right) Thermostatically controlled propagators are very versatile.
(Far right) An electric winter heater assists the cultivation of more exotic crops.
(Centre right) Soil warming cables heat large areas economically.
(Right) Heat mats offer gentle bottom heat.

Propagators

Having owned the same propagator since my late teens, I can now (30 years later) say that this heated model with thermostatic control was a wise investment. Every year I tweak the warmth according to its contents; 24°C (75°F) in winter for chillies and tomatoes, 18°C (64°F) come spring for squashes and French beans. Current models offer ultimate control with integrated lighting and water reservoirs. Unheated propagators are less expensive and ideal for indoor use, including perfectly formed windowsill models.

Heat mats

These electrically heated mats are versatile and space-savvy. They can be placed underneath tender plants to get them through winter. There is no need to jostle pots around to fit, unlike with sealed propagators. Cover plants with clear plastic lids to keep them moist. Crudely adjust heat using layers of cardboard, or more accurately with a thermostat.

Heat cables

These come in various lengths, and inexpensively warm a larger area. Bury them 10cm (4in) deep in greenhouse borders for winter cropping – excellent for hardy salad leaves – or snake into a shallow case of sand, as a bespoke propagator or overwintering vessel. If you don't want them on continuously, install a timer or thermostat.

Heaters for winter

Heating a whole greenhouse in winter is a luxury, but there are efficiency tweaks to make it affordable. Heating just part of the structure is most economical by creating a warm insulated "zone". With warm air rising, position the heater low. Bear in mind, too, that electric heaters are far cleaner than gas or paraffin (the latter duo give off water and fumes, whereas electricity produces no by-product). Timers are often less costly than thermostats, but offer less accuracy.

TWEAKS TO OPTIMIZE YOUR PLOT

Having a good armoury of materials on the plot means that you're ever-ready for scrambling crops, nuisance pests, and sudden chills. Many, with care, can be used year on year. Some even have the potential to be grown by yourself, saving on budget and ticking the sustainable box. Ensuring you are fully equipped with these essentials ensures that growing your own food remains a pleasure.

Supports, barriers, and insulation

Grow, build, or weave your own

Home-grown supports

You may already have plants perfect for supports. Bamboo, hazel, birch, dogwoods, and willows, all work. I remove branches from my five hazel trees each year for pea sticks and poles. The pea sticks are excellent for broad bean cages and wigwams for smaller climbing edibles; the poles for climbing beans, cordon tomatoes, and squashes. Trees double up as windbreaks (see pages 122–123). Stack logs from larger trees post-surgery, in a shady, damp area as a beetle habitat, or create a mushroom "loggery" (see pages 154–155).

Posts and wires

Wires on a fence, wall, or supporting posts, are invaluable for fruit cultivation. Use them to train fruit into intricate shapes. Fan-trained fruit has a tiny footprint compared to yield; a system of wires and straining bolts utilizes all available growing space (see pages 74–75). Rows of free-standing plants like raspberries or cordon apples make useful boundaries. Erect sturdy posts for longevity.

TWEAKS TO OPTIMIZE YOUR PLOT

Espalier apples (above far left) and cordon tomatoes (top right) appreciate sturdy supports. You can grow your own supports of dogwood (above), hazel and birch. Twine, beanpoles and pea sticks are all useful as supports for Malabar spinach (far left), lablab beans (centre left), and broad beans (below left). Vine eyes and straining bolts (above left) help keep wires tight. (Top, middle) leek moth can easily be prevented using fine insect-proof mesh.

Pest control

Netting and mesh are invaluable pesticide-free pest control. Own three sizes: 2cm (¾in) for protecting fruit canes and bushes from birds; 5mm (¼in) for protecting brassicas from butterflies; and 0.8mm (1/16in) for tiny insects like leek moth, cabbage root fly, and carrot fly. Wire bird and butterfly mesh last years; starch-based, biodegradable insect meshes are available.

Crop insulation

Insulating materials are essential for growing global edibles outdoors. Overwinter borderline hardy crops such as pomegranate, feijoa, and loquat in sheltered pots or well-drained beds. Create a simple cage from bamboo, then add bubble wrap (recycled and reusable) as an excellent water-repellent lid. Hessian is a great alternative to horticultural fleece for lining cage sides – both encourage good airflow, deterring winter rot. Place insulating straw inside cages for added frost protection.

Crop covers, large or small, help to nurture and protect edibles. A glasshouse or polytunnel – the ultimate structure – offers a variety of uses. Smaller cloches and frames also vary in what they can offer. All are invaluable when growing your own food.

Covers

Glasshouses, cloches, tunnels, and frames

Greenhouses

A greenhouse for spring propagation is ideal. Glass is very insulating, as is twin-walled polycarbonate, and its lifespan significant. One in full sun can extend your growing season from very early spring, well into autumn – heat regulation can prove tricky in summer, however. A greenhouse in part shade still offers good light and heat retention for spring and autumn cropping. Though chillier in winter, summer heat will be easier to manage. Aim for lower and upper vents to encourage optimum airflow.

Polytunnels

Polytunnels are often less expensive than glasshouses, so are great for larger plots. Consider polythene has a shorter lifespan and is less insulating than glass. Ample ventilation is key – models with large adjustable side vents are ideal. The same location principles apply as for greenhouses.

(Top left) Large covers can cloak whole raised planters, with zips (top mid) offering easy access. (Top right) Frames are useful for overwintering Mediterranean herbs. (Right) Greenhouses are excellent for propagation. (Far right) Low glass cloches are sturdy and effective.

TWEAKS TO OPTIMIZE YOUR PLOT

Frames

Larger rigid frames constructed from metal/wood and glass or polycarbonate can be lined with polystyrene to overwinter half hardy edibles, such as pomegranate and feijoa – position in full sun if possible. Stackable timber pallet "collars" are also useful for this. Smaller frames are perfect for overwintering herbs that dislike winter wet, such as tarragon and lemon verbena. They are also great for overwintering broad bean seedlings and hardwood fruit cuttings. Mini, upright "growhouses" with polythene covers take up little floor space. Line the sides with insulating materials to offer winter frost protection.

Cloches

Mesh tunnel cloches make useful pest control for rows of edibles; rigid plastic or glass ones are excellent for forcing early crops in a sheltered, sunny spot. Rows of leafy hardy veg, such as spinach and chard, appreciate a polythene tunnel cloche in cold climates to keep foliage palatable. Secure lightweight cloches well in exposed sites.

While composting and mulching are both activities that recycle nutrients back into the garden, additional fertilizer support is beneficial when gardeners are harvesting, pruning, and aiming for high yields. Use the right feeds in targeted applications for efficiency – you can even grow your own.

Nutrition

Source the right feeds and use them responsibly

Essential fertilizers

The rows of products promising fantastic results on garden centre shelves can feel bewildering. In my experience, you only need a few. You want to encourage fruits (for example strawberries, using potash), leaves (spinach, using nitrogen), roots (carrots, using phosphorus), or a mixture of all three (transplants and young crops). Supply these nutrients as a liquid that plant roots can quickly take up, or slowly over a whole season – usually granular.

Fish, blood, and bone is an organic, animal-derived feed that slowly releases all three major nutrients, ideal for long-term crops like fruit trees, bushes, and canes. Organic liquid balanced feeds offer a quick-release alternative. To boost flower and fruit production, potash is useful. For a slower-release formulation comfrey pellets work well, and for quick-release, use an organic liquid tomato feed.

Targeting and timing

Excess feed will not give you bigger harvests and can scorch plants, or seep into the groundwater causing problems. Always apply the dosage recommended and only feed during active growth – between mid-spring and early autumn. Don't feed a whole bed unless necessary; apply slow-release organic feeds into the soil just above the root zone, and liquid feeds to plant bases, not bare soil.

Grow your own feed

Grow comfrey as a source of a balanced liquid feed – cut the leaves, rot them down in water, and use the resulting liquor. The variety 'Bocking 14' is nutritious and least likely to self-seed. Sow "green manures" like clover, mustard, and ryegrass in autumn on bare vegetable beds. Dig in during autumn or spring to release their nutrients for subsequent crops.

(Above) Powdered or granular organic feeds offer a slower release of nutrients. (Centre right) Sow green manure crops on bare soil to "catch" nutrients. (Below right) Comfrey is a useful nutrient accumulator.

TWEAKS TO OPTIMIZE YOUR PLOT 201

(Top left) Liquid feeds are quickly taken up by crop roots. (Top right) Harvest comfrey leaves in midsummer to make a potash feed. (Above) *Phacelia* is a good bee attractant, as well as a green manure.

Recycling green waste is applaudable. It reduces landfill volume and methane production, and yields an invaluable material teaming with life – your crops will love it. There are systems available for every plot size – or no plot at all. Embrace the process, to create a soil conditioner, potting compost activator, mulch, and liquid fertilizer.

Composting

Systems for plots of all shapes and sizes

Different systems

If you want to produce significant volumes of compost, a three-bay system is perfect. One bin is maturing, the second being filled, and the third free to turn this material into. (Do this every 2 or 3 months.) One cubic metre (35 cubic feet) or larger generates sufficient heat for speedy rotting and killing off weed seeds. Sealed, smaller units such as hot bins, tumblers, and "dalek" bins are less obtrusive, each promising their own merits. Turn these every 3 months too. Smaller still are wormeries that harness the appetite of tiger worms, recycling garden waste (but not all items). These are ideal for courtyard and balcony gardens. Kitchen worktop systems like bokashi fermenting units are great for collecting domestic food waste.

What to add

Add the vast majority of garden waste to your compost, chopping it up beforehand to speed up decomposition. Exceptions include weed seedheads or root fragments of perennial weeds, and debris carrying soilborne pests or diseases like root fly or clubroot. Adding these items encourages such problems to manifest and spread around your plot.

Processing your compost

Once fully rotted, dig compost into soil to improve soil texture, improving drainage in clay soils, and moisture retention in sandy ones. You can also use compost as a mulch. Garden compost often contains woody lumps. Pass material through a garden shredder before composting, or through a coarse sieve like chicken wire once rotted, to make a more attractive mulch or potting compost additive. Wormery and bokashi waste is very rich, and best used as a soil improver, or for adding to potting compost in small amounts. Use wormery run off as a general liquid feed – diluted 1 part liquid to 10 parts water.

Producing your own compost yields a nutrient-rich product for your plot. A three-bay system (bottom left) is the ultimate luxury. Wormeries are ideal for compact plots (left) and "dalek" bins (bottom right) are commonplace on allotments. Sieving homemade compost (below) eliminates any coarse woody lumps for a finer product.

Troubleshooter

Troubleshooter

Become finely tuned to your plot and you'll gain insight into what creatures you share it with. Prevention of problems is frequently possible once you know what might appear year on year.

Allium leaf mining fly
Cloak susceptible plants (leeks, chives, onions, shallots, garlic) in fine 0.8mm (1/16in) insect-proof mesh, holding mesh away from crop foliage with hoops. If affected, rotate alliums to clean soil for 1 year.

Annual weeds
Regularly hoe any patches of bare soil so that annual weed seedlings cannot establish, or lay thick 4–5cm (1½–2in) organic mulches over bare areas. Be tenacious in removing any annual weeds in flower or setting seed.

Aphids on broad beans
Monitor crops regularly from spring onwards, squashing colonies between finger and thumb before they establish, or squirting them with a strong jet of water. Repeat this process frequently to keep numbers to an acceptable level.

Apple canker
Remedy wet, waterlogged, or acidic soils before planting apples on them, improving drainage and/or liming. Prune out affected shoots well into healthy growth. Choose resistant varieties (e.g. 'Katy', 'Winston').

Asparagus beetle
Monitor crops regularly, and hand-pick off adult beetles or larvae when seen (squash egg colonies between finger and thumb). Biological control nematodes can be used. Take old foliage off-site in autumn.

Allium leaf miner

Aphids on vegetable crops

Apple canker

Annual weeds

Asparagus beetle

Bacterial canker

Prune vulnerable trees (all fruiting *Prunus* crops) in July or August, immediately after harvest (young trees yet to fruit well can also be pruned in May or June) – cut out obvious cankers. Some varieties show resistance.

Bean rusts

If known to be a problem locally, grow vulnerable plants (runner and French beans, plus broad beans) in an open, airy site. Thin out leaf growth to reduce humidity, and avoid overhead watering. Clear up all debris and don't save seeds.

Birds

Net vulnerable crops (pigeons feed on brassicas, blackbirds and thrushes adore fruit) when they begin to ripen. Do NOT hang netting loosely on plants, birds can get caught – pin it securely and taut to frames to encage crops.

Cabbage root fly

Place "collars" (15cm/6in-diameter cardboard circles) around transplants, or grow vulnerable plants (brassicas) under insect-proof mesh. If soil is known to be infected, rotate brassicas to a clean area.

Carrot fly

Grow vulnerable crops (carrots, parsnips, parsley, celery, celeriac) under fine, insect-proof mesh. If damage is seen, rotate susceptible crops to a new bed for a year. Some carrot varieties (e.g. 'Resistafly' and 'Maestro') show resistance.

Bacterial canker

Bean rusts

Birds

Cabbage root fly

Carrot fly

Club root

Improve drainage and lime soils to raise pH to 8.0. In infected soil, grow brassica seedlings individually in 1-litre (4–5in) pots before planting out for strong roots. Choose resistant varieties (e.g. cauliflower 'Clapton', swede 'Marian').

Codling moth

Hang pheromone traps in vulnerable trees (apples and pears) in May, to trap male moths. Biological nematode controls can be used in autumn against caterpillars leaving the fruits. Encourage birds, beetles, and hedgehogs onto your plot.

Cutworms

Grow affected edibles (lettuces, root, tuber crops) under insect-proof mesh to minimize damage. Disturbing soil exposes these caterpillars to natural predators. Apply biological nematode controls, if damage cannot be tolerated.

Deer

Place spiral guard around vulnerable trees, or erect 6ft (1.8m) tall livestock fencing around areas to be protected. Dog, human scent, or other repellant substances may deter deer, but need frequent reapplication.

Downy mildew

Avoid planting densely, instead encourage good air flow with wide spacings. Water in the morning so that excess water quickly evaporates off foliage. Remove affected foliage promptly. Grow resistant crops (e.g. lettuce 'Lettony').

Club root

Deer

Codling moth

Cutworm

Downy mildew

TROUBLESHOOTER

Flea beetle
Grow vulnerable plants (brassicas) under insect-proof mesh, transplanting them out as good-sized plants that can quickly grow to a less susceptible size.

Fruit aphids
Monitor crops in spring, squashing colonies between finger and thumb. Encourage natural enemies, which by late spring will have normally reduced aphid clonies to an acceptable level.

Gooseberry sawflies
Monitor plants from April onwards, for larvae activity, picking off all that are seen. Biological nematode controls can also be applied when the larvae is present. Hoe under plants in summer to disturb pupae and expose them to predators.

Honey fungus
Remove affected plants (along with their root system) promptly. Sink vertical barriers 45cm (18in) deep to control spread, changing soil if needing to replant in an affected area. Grow resistant crops (e.g. *Gaultheria*).

Late blight
Patrol vulnerable crops (tomatoes and potatoes) in warm, wet weather, removing affected plants promptly. Defoliate affected potatoes to save the tubers. Grow resistant varieties (e.g. tomato 'Crimson Crush', potato 'Sarpo Mira').

Flea beetle

Honey fungus

Fruit aphids

Gooseberry sawfly

Late blight

Leaf miner

Vulnerable plants (spinach, chard, beetroot, celery, celeriac, parsley) can be grown under fine, insect-proof mesh. Pick off light damage.

Leek rust

Avoid planting densely, instead encourage good air flow with wide spacings. Grow autumn-planted garlic and lift before it succumbs. Don't over-apply high-nitrogen feeds. Grow resistant varieties (e.g. leek 'Below Zero').

Mice and voles

Fit young trees with spiral guards if bark gnawing is noticed (you will see telltale tiny teeth marks). Grow seedlings on in an area where mice do not have access before planting out. Make stores rodent-proof.

Onion white rot

If the problem exists on your plot, grow vulnerable crops (onions, shallots, leeks, chives, and garlic) in 1-litre (4–5in) pots before planting out. Eat crops fresh rather than storing them (they can be chopped and frozen).

Pea moth

Sow early peas under glass in spring, to plant out and crop before moth populations rise (eggs are laid in June and July). Grow plants under insect-proof mesh during egg-laying times. Sow a late-maturing (August) crop.

Leaf miner

Onion white rot

Leek rust

Mice

Pea moth

Peach leaf curl

Cloak vulnerable trees (peaches and nectarines) under clear polythene sheeting between late January and late April – this is far easier on fan-trained plants. Hand-pollinate flowers on such trees to ensure fruit set.

Pear midge

Monitor trees for affected fruitlets as they form from April onwards, removing any seen before the midge has time to fall to the soil to pupate. Hoeing under infected trees in summer exposes some pupae to dehydration and natural predators.

Pear rust

Try to tolerate damage, but if it seems detrimental, remove heavily affected leaves. Junipers are alternate hosts but their removal doesn't guarantee control. Cankers are uncommon but if seen should be pruned out.

Perennial weeds

Grow only annual crops in areas colonized by perennial weeds, so that they can be cleared periodically and dug over to remove all fragments of weeds. Sink vertical barriers alongside potential areas of re-infection.

Plum moth

Hang pheromone traps in vulnerable trees (plums, gages, damsons) in May, for male moths. Apply biological control nematodes in September for larvae leaving the fruit. Encourage birds, predatory beetles, and hedgehogs.

Peach leaf curl

Pear midge

Perennial weeds

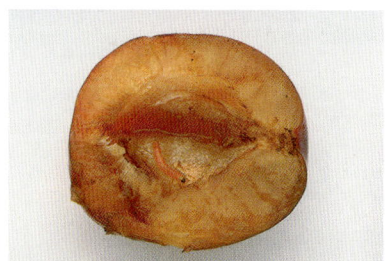
Plum moth

Pear rust

Powdery mildew

Avoid planting densely, instead encourage good air flow with wide spacings and prune to thin canopies. Deter drought stress (but limit overhead watering). Grow resistant varieties (e.g. gooseberry 'Invicta', courgette 'Defender').

Rabbits

Place spiral guards around young trees. Alternatively install livestock fencing, 1.2m (4ft) high, buried 30cm (12in) into the soil (bend this base outwards to deter burrowing). Animal repellent substance may have some effect but require frequent re-application.

Raspberry beetle

Hoe under affected plants in summer to expose pupae to predators. Hang host plant odour (karimone) traps in mid-spring. Grow later-maturing, less affected varieties of vulnerable plants (raspberries, blackberries, and hybrid berries).

Replant disease

Change the soil in the affected area for some that hasn't supported fruit trees for many years (e.g. vegetable garden soil). Apply mycorrhizal root inoculants when planting vulnerable plants, and choose rootstocks with resistance.

Slugs and snails

Encourage natural predators such as thrushes, frogs, and predatory beetles by providing suitable habitats. Night patrols can help dent heavy populations, especially if they have built up in warm, wet conditions.

Powdery mildew

Raspberry beetle

Rabbits

Replant disease

Slugs and snails

Spotted wing drosophila

When fruits are full-sized but before they show signs of colour, cloak individual branches or whole trees in insect-proof mesh (this is far easier if your fruit trees, canes, and bushes are trained against a wall or fence).

Squirrels

Strong wire netting is the most effective way to protect edibles from damage by squirrels. Make up frames of it, binding them together over plants when vulnerable (often when fruit is swelling). Wall-trained trees are easier to cover.

Vine weevil

Monitor vulnerable plants (includes many fruits, especially potted) and encourage natural predators such as birds, frogs, rodents, hedgehogs, and beetles. Dispose of affected compost. Consider biological nematode controls in early autumn.

Viruses

Remove affected plants promptly and disinfect any tools used. Control sap sucking insects as many of these transmit viruses. Sow seeds of virus-resistant varieties, and buy certified virus-free plants.

Woolly aphid

Small colonies can be pruned off, if on non-essential twigs and small stems. On larger limbs and trunks use a stiff bristled brush to dislodge colonies.

Spotted wing drosophila

Viruses

Squirrels

Apple woolly aphids

Vine weevil

Glossary

Acid soil – soil with a pH value below 7

Alkaline soil – soil with a pH value above 7

Annual – A plant that completes its life cycle in one year

Biennial – A plant that completes its life cycle in two years

Biological control – Controlling pests by introducing natural parasites or predators

Blanching – Covering a plant to reduce the amount of light it receives, often resulting in more palatable growth

Bolting – When a crop prematurely produces a flower spike

Brassicas – A family of plants including cabbages, broccoli, and turnips

Brix reading – A measurement of dissolved sugars in a given plant liquid

Bud – An immature organ enclosing an embryonic leaf, stem, or flower

Chlorophyll – A green pigment that absorbs energy from the sun

Carbon farming – Managing land to maximize the amount of carbon stored

Chitting (potatoes) – Encouraging tubers to produce shoots before planting

Chlorosis – Leaves yellowing due to a lack of chlorophyll

Cloche – A small portable structure used to protect early crops

Cold frame – A large, glazed box used to protect plants from excess cold

Cordon – A fruit tree restricted to (usually) one main vertical or diagonal stem

Cultivar – a cultivated variant of a plant species.

Cut-and-come-again – Repeatedly harvesting leaves to prolong the picking period

Damping down – Drenching glasshouses or polytunnels with water to boost humidity

Direct-sow – Sowing seeds outside directly into the soil

Drill – A line of seeds or seedlings

Espalier – A fruit tree with horizontal left and right tiers, growing from one main vertical trunk

F1 hybrid – Vigorous offspring obtained by crossing two selected parent plants

Fan – A fruit tree with multiple branches radiating out from a low central point

Foliar feed – A feed applied to, and taken up via, the leaves of plants

Forcing – To encourage early plant growth (usually) by applying heat

Grafting – To artificially join one or more plant parts to another

Green manure – A quick-growing leafy crop that is dug into the soil to add nutrients

Half-hardy – A plant that can withstand cold temperatures but dies when frosted

Harden off – Gradually acclimatizing indoor-raised plants to life outside

Hardy – A plant able to withstand cold temperatures, including freezing

Heeling in – To temporarily plant until the crop can be placed in its final position

Hungry gap – A period in spring when little fresh produce is available

GLOSSARY

Hydroponics – Growing plants in dilute solutions of nutrients

June drop – When a fruiting plant naturally sheds some of its fruitlets in early summer

Legginess – Seedlings with excessive stem length, usually because of lack of light

Legume – A family of plants including peas and beans

Mulch – Material added to the soil surface to retain moisture and suppress weeds

Neutral – soil with a pH value of 7

Organic matter – Mulch, compost, or similar material derived from plant materials

Osmosis – The diffusion of water molecules from a dilute solution to a more concentrated solution across a selectively permeable membrane

Pea sticks – Twiggy, long stems used to train peas, beans, and other edibles

Perennial – A plant that lives for several years

Pheromone trap – A trap that uses pheromones (naturally occurring chemicals) to attract insects.

Photosynthesis – How plants use sunlight to synthesize nutrients from carbon dioxide and water

Pollination – The transfer of pollen from floral anthers (male) to stigmas (female)

Pricking out – Moving seedlings from where they have germinated into individual pots or beds

Rootball – The roots and accompanying soil/compost of a (woody) plant

Rootstock – A plant used to provide the root system for a grafted plant

Sap bleeding – The release of sap from a (woody) plant via a wound or pruning cut

Seep hose – A porous hosepipe that slowly releases water along its length

Sessile – Permanently attached to the soil, immobile

Shelterbelts – A barrier of plants that provides protection from wind

Sideshoot – A stem that arises from the side of a main stem

Soil conditioner – Materials added to soil to improve aeration, water retention, and fertility

Spur – A short branch on woody plants bearing flower buds

Spur pruning – Cuts made (usually) to sideshoots to encourage spurs to form

Station-sow – To sow seeds in small clusters at regular intervals along a row

Tender – A plant that is vulnerable to, and will be killed by, frost

Thin (seedlings) – The removal of a proportion of seedling plants to improve the growth of those that remain

Threshing – Removing seeds from the husk and stems that carry them

Top dressing – Applying a material (e.g. powdered fertilizer) to the soil surface

Transplant – Moving a plant from one position to another

Variety – A (usually) naturally occurring variant of a plant species

Vine eyes – A holed metal fixing, used to secure wire to walls and fences

Useful resources

Seeds and/or vegetable plants

Mr Fothergill's
Phone: +44 (0)333 777 3936
Website: www.mr-fothergills.co.uk

Chiltern Seeds
Phone: +44 (0)1491 824675
Website: www.chilternseeds.co.uk

Real Seeds
Phone: +44 (0)1239 821107
Website: www.realseeds.co.uk

Incredible Vegetables
Website: www.incrediblevegetables.co.uk

Sea Spring seeds
Phone: +44 (0)1308 897898
Website: www.seaspringseeds.co.uk

> Buying seeds, bulbs, tubers and plants:
> Always buy plants from established and trusted British mail order or internet sellers, to help protect the UK from the introduction of unwanted pests or diseases.

Fruiting plants

Jurassic Plants
Phone: +44 (0)7909 100255
Website: www.jurassicplants.co.uk

Agroforestry Research Trust
Phone: +44 (0)1803 840776
Website: www.agroforestry.co.uk

Mushroom kits

Caley Bros
Website: www.caleybrothers.co.uk

Self-watering pot systems

AutoPot
Website: www.autopot.co.uk

Information

Plants For A Future
Website: www.pfaf.org

Inspiring edible gardens

RHS Garden Hyde Hall
Phone: +44 (0)1245 400256
Website: www.rhs.org.uk/gardens/hyde-hall

RHS Garden Wisley's World Food Garden
Website: www.rhs.org.uk/gardens/wisley/garden-highlights/the-world-food-garden
Phone: +44 (0)1483 224234

Index

Main entries are in **bold**

A
achocha 32, **73**
acid soils 57
agaves 17, 32, **182**
agretti 10, 27, 37, 40, **83**
alkaline soils 57
allium leaf mining fly 206
allotments 48–49
aloe 184
Alpine strawberries 16, 20, 21, **160**
altitude 37
amaranth 26, 27, 40, **87**
aphids 206, 209, 213
apple canker 206
apples 19, 30, 31, **118**, 197
apricots 16, 18, 39, **62**, 74
arbutus 39
Arctic raspberry 120
artificial light 192
artificial light growhouse 186–87
Asian broccolis **96**
asparagus 10, 27, 40, **78**
asparagus bean 72

asparagus beetle 206
aubergines 10, 33, **70**
avocados 32

B
bacterial canker 207
balconies 16
bamboo 17, 21, **167**
Barbados aloe 184
barberries **169**
basil **64**
bay 38, 39, **73**
bean rusts 207
beans
    asparagus bean 72
    broad beans 26, 36, 38, **112**, 197
    drying beans **95**
    French beans **95**
    hyacinth bean 104
    lima beans 10, **81**
    runner beans 40, **65**
    soybeans 32
beetroot 30, 57, **114**
berry patches 138–39
bilberries 21
birds 207
blackberries 21, **168**

black chokeberry 152
blackcurrants **145**
blight 209
blueberries 24, 25, 31, 57, **115**
boglands 25
bokashi 201
bolting 25, 30
brighteyes 136
broad beans 26, 36, 38, **112**, 197
broccoli **98**
    Asian broccolis **96**
Brussels sprouts 31, **119**
bulbous nettle **153**
bullaces 113
butternut squashes 18, 19

C
cabbage root fly 207
cabbages 31, 38, 57, **114**
cacti 32
calcium 57
candyleaf 185
canker 207
cape gooseberry 73
cardamom 180
carrot fly 207
carrots 31, 38, **111**
case studies 42–49

Caucasian spinach 10, **152**
cauliflowers **119**
celeriac 10, 21, **142**
celery 9, 21, **142**
chalk soil **54–55**, 57
chard 11, 17, **132**, 185
chayotes 32
cherries **63**, 74
chervil 16, 25, **153**
chicory 120, 185
Chilean guava 131
chillies 16, 33, 38, 39, **69**, 183
Chinese artichoke **153**
Chinese flowering quince 119
chokeberries 152
choy sum **96**
cinnamon vine 72
citrus fruits **179**
city gardens and heat 32
clay soil 53, **54–55**, 56
cloches 199
club root 208
codling moth 208
cold areas 30–31, 34–35
comfrey 200
composting 201–02
coriander 136
courgettes 25, **97**
courtyards 16
covers 198–99
cranberries 24, 25
crowberries 21
cucumber 25, **105**
cutworms 208

D
dahlias 19, **101**
damsons 113
deer 208
dill 105
diseases 206–13
    soil pH 57
diversity 10
dog rose 120
dog's tooth violet 149
downy mildew 208
drainage 25, 26
drosophila, spotted wing 213
drought-resistance 26
dry areas 26–27, 28–29
durum wheat 32

E
elder 129
endive 120, 185

F
false shamrock 184
fan training 74–75, 196
feijoa 11, 26, 38
fennel 10
    Florence fennel 25, **103**
fertilizers
    essential 200
    reducing usage 10
figs 9, 16, 18, 26, 33, 74, **82**
flavour, sun effects 18
flea beetle 209
Florence fennel 25, **103**
frames 199
freezing crops 19
French beans **95**
frost pockets 35
fruit, training 18, **74–75**, 196
fuchsia 136

G
gages 40, 74
galangal 180
garlic 39, **88**
garlic cress 137
garlic mustard 21, **164**
garlic, wild **168**
gaultheria 17, 24, 39, **146**
ginger 180
globe artichoke 10, 26, 36, **86**
gluts 10
gooseberries 16, 21, 74, **158**
gooseberry sawflies 209
goosefoots 84
grapes 17, 18, 26, **85**
greenhouses 34, 194, 198
green manures 200
growhouse, artificial light 186–87
guava 131

H
hardiness 19, 31
hazelnuts **162**
heart's ease 152
heaters and heat mats 194
hedging 40
herbs
    dry shade 21
    water loving 25
    windowboxes 16

hibiscus 185
high altitude 37
honeyberries 21, **168**
honey fungus 209
horn of plenty 10, **89**
horseradish **161**
hostas 10, 20, **148**
hot areas 32–33, 34–35
hyacinth bean 104

H
Indian shot 104
indoor crops 173–85
indoor growing 192
insulation 197
iron 57
irrigation 190

J
Japanese wineberries 10, **163**
Jerusalem artichokes 26, 36, **89**
jostaberries 21, **168**

K
kale 30, 31, 36, **114**
kitchen gardens 42–43
kiwis 39, **73**
kohlrabi 117
komatsuna 137
Korean celery 120
kumquats **179**

L
land cress 16, 25, **169**
leaf miner 210

leek rust 210
leeks 30, 31, 36, **116**, 197
lemongrass 16, **178**
lemons **179**
lemon verbena 185
lettuces 30, 37, **100**, 185
light usage 192–93
lima beans 10, **81**
limes **179**
lingonberries 21, 24, 36
loam **54–55**
loganberries 21, **168**
loquat 26
lovage 137

M
magnesium 57
Malabar spinach 32, 33, **68**, 196
mapping your plot
    cold and hot 34–35
    overview 16
    sun and shade 22–23
    wet and dry 28–29
    wind and shelter 40–41
mashua 21, 136
melons 38, **104**
mibuna 150
mice 210
microclimates, mapping 16
microleaves 176
mildew 208, 212
mint 130
minutina 25, **151**
mitsuba 25, **153**
mizuna 25, **150**, 185

mooli 25, **104**
mulch 26, **170–71**, 190
mushroom 'loggery' 154–55
mushroom plant 184
mustards 119, 185

N
nasturtiums 137
nature, working with 10
nectarines 18, 38, **62**, 74
nettles 21, **153**, **166**
New Zealand spinach 10, **88–89**
nitrogen 56, 57, 200
nutrition 56, **200–201**

O
oca 88
onions 17, 36, 39, **110**
Welsh onion 136
onion white rot 210
open and cold zones
    crops 110–21
    overview 109
opium poppy 89
orache 10, **88**
oregano 16, **88**
Oregon grapes **169**

P
pak choi **134**, 185
parsley 21, 25, **147**
parsnips 31, **111**
patience dock **165**
patios 16
peaches 16, 18, 19, 38, **62**, 74

peach leaf curl 211
pea moth 210
pear midge 211
pear rust 211
pears 19, 41, **102**
peas 19, **119**
pea shoots **175**
peat **54–55**, 57
pelargoniums 185
peppers **69**
perilla 11
pest control 197
pests 206–13
pH of soil 57
phosphorus 56, 200
pineapple guava 72
pinkcurrants 74, **159**
plum moth 211
plums 74, **113**
plum yews **169**
polytunnels 198
pomegranates 26, 38, **181**
potash 200
potassium 56
potatoes 38, **99**
pots, growing in 16
powdery mildew 212
prickly heath **146**
prickly pear 32, **177**
propagation, indoor 192
propagators 194
pumpkins 10, **94**
purslane 26

Q
quinces 72, 119
quinoa 87

R
rabbits 212
radicchio 120
rainfall capture 190
rain shadows 29
raised beds 25, 29
raspberries 21, 31, **128**
Arctic raspberry 120
raspberry beetle 212
redcurrants 9, 21, 74, **159**
red orache 88
replant disease 212
resources, conservation 10
rhubarb 16, 21, 31, **143**
ripening, after cropping 19
rocket 25, 30, **126**, 185
Turkish rocket 89
rosehips 120
roselle hibiscus 185
rosemary 36
runner beans 40, **65**

S
sage 16
salad rocket 126
salads, indoor growing 185
salsify 10, 27, 36, **79**
sandy soil 52, 53, **54–55**, 56, 57
Savoy cabbage 31
scorzonera 26, **79**
sea beet 9
sea buckthorn 11, **119**
seakale 10, 36, **89**
serviceberry 137
shade 20–21, 22–23
shade, partial, zones
    crops 126–37
    overview 125
shady and dry zones
    crops 158–69
    overview 157
shady and wet zones
    crops 142–53
    overview 141
shallots 37, **110**
shelter 37, 38–39, 40–41
shelterbelts 37
shiso 105
shuttlecock ferns 10, 20, 152
sloes 113
slugs 212
snails 212
soakaways 25, 56
soil 52–57
solar power 192
Solomon's seal 10
sorghum 32
sorrel 10, 16, **153**
soybeans 32
spinach
    annual 10, 25, 30, **127**
    Caucasian 10, **152**
    indoor growing 185
    Malabar 10, 32, 33, **68**, 196

INDEX

New Zealand 10, **88–89**
perpetual 132
tree spinach 10, 26
spotted wing drosophila 213
spring onions 105
squashes 10, 17–19, 24, 39, **94**, **97**
squirrels 213
storing crops 19
strawberries 26, 39, **66**
    Alpine 16, 20, 21, **160**
summer purslane 10, 37, 40, **80**
summer squashes **97**
sun 18–19, 22–23
sunflowers 32
sunny and moist zones
    crops 94–105
    overview 93
sunny and sheltered zones
    crops 62–73
    overview 61
sunny, open, and dry zones
    crops 78–89
    overview 77
supports for plants 90–91, 196
swede 117
sweet cicely 21, **168**
sweetcorn 19, 32, 36, **104**
sweet potatoes 19, **71**
Swiss chard 11, 17, **132**
Szechuan pepper 135

T
tatsoi 134
tayberries 168
tea 184
temperature management 194
Texsel greens 133
thermophiles 32
thyme 16, 37
tomatillos 39, **73**
tomatoes 10, 16, 17, 18, 19, 38, **67**, **174**, 197
topsoil 53
training fruit 18, 74–75, 196
trees, effect on soil moisture 29
tree spinach 10, 26
Turkish rocket 89
turmeric 180
turnips 38, **117**

U
udo **169**
urban heat islands 32

V
vernalization 30
vine weevil 213
viruses 213
voles 210

W
wasabi 9, 24, **153**
watercress 24
waterlogging 56
water management 190–91
weeds
    annual 206
    perennial 211
    prevention 106–07
Welsh onion 136
wet areas 24–25, 28–29
wheatgrass 184
whitecurrants 74, **159**
wild garlic **168**
wind 36–37, 40–41, 90–91
    see also open zones
windbreaks 37, **121–22**
windowboxes 16
wind tunnels 41
wineberries 21, **163**
winter purslane 25, **144**
winter squashes **94**
woolly aphid 213
worcesterberries 21, **168**
wormeries 201–02

Y
yacon 19, **105**
yams 19

About the author

Lucy Chamberlain is the daughter of smallholders who sold produce to the London food markets for 30 years, via their 5-hectare (12-acre) Essex nursery. This ignited her passion for horticulture, and for growing all manner of edibles. After studying horticulture at college, she worked as a Horticultural Advisor for the Royal Horticultural Society, then edited *Grow Your Own* magazine, before being appointed as Head Gardener on an 35-hectare (85-acre) Essex estate with a large walled kitchen garden. Lucy authored the *RHS Step-by-Step Veg Patch*, which to date has sold well over 100,000 copies. She now works as a freelance writer, gardener, podcaster, and broadcaster, busying herself within her 12 × 12m (39 × 39ft) modern-day kitchen garden where she grows more than 150 varieties of fruits, vegetables, and herbs.

Acknowledgements

The author would like to thank

All at Dorling Kindersley and the Royal Horticultural Society, especially Chris Young, Ruth O'Rourke, Helen Griffin, Barbara Zuniga, Christine Keilty, Dominique Page, Alice McKeever, Sophie Blackman, Good Wives and Warriors, and Andrew Torrens, for believing in my idea and helping to bring it to fruition.

Neil Hepworth for his wonderful photography. Maz for the loan of his drone, and garden owners Simon, Chris, Mike, and Maureen and Tony for their kind permissions.

Jurassic Plants, Agroforestry Research Trust, Chiltern Seeds, Real Seeds, Mr Fothergill's, AutoPot and Caley Bros for supplying props, seeds, and plants for photoshoots.

Guy Barter and Matt Oliver for guidance on soils and fuchsias.

All staff at RHS Hyde Hall (especially Robert Brett and Tom Freeman) for garden photography.

Ian for helping to prepare the garden for photoshoots.

My family for eating various edibles over the years in the name of research.

The publisher would like to thank

Nicola Hodgson and Alastair Laing for editorial support.

Picture credits

The publisher would like to thank the following for their kind permission to reproduce their photographs:

(Key: a–above; b–below/bottom; c–centre; f–far; l–left; r–right; t–top)

138 Dreamstime.com: EMFA16 (c). **140 Dreamstime.com:** Volodymyr Pishchanyi (ca). **154 Dreamstime.com:** Ludovikus (tr). **206 Dreamstime.com:** Maumyhata (bl); Tom Meaker (br). **207 Dreamstime.com:** Tomasz Klejdysz (cr); Paul Maguire (cb). **208 Dreamstime.com:** Andrei Shupilo (bc); Abdelmoumen Taoutaou (cr). **211 GAP Photos:** Claire Higgins (c). **212 Dreamstime.com:** Tomasz Klejdysz (tr). **GAP Photos**: Andrea Jones (cr). **213 Dreamstime.com:** Brett Hondow (cr); Tomasz Klejdysz (tc); Kewuwu (br). **GAP Photos:** Dave Bevan (tr)

All other images © Dorling Kindersley Limited

Editorial Manager Ruth O'Rourke
Senior Editor Sophie Blackman
Senior Designer Barbara Zuniga
Production Editor David Almond
Senior Production Controller Samantha Cross
DTP and Design Coordinator Heather Blagden
Art Director Maxine Pedliham
Publishing Director Katie Cowan

Editorial Alice McKeever, Dominique Page
Design Christine Keilty
Photography Neil Hepworth
Illustration Andrew Torrens
Jacket illustration Good Wives and Warriors
Consultant gardening publisher Chris Young

ROYAL HORTICULTURAL SOCIETY
Consultant Simon Maughan
Books Publisher Helen Griffin
Head of Editorial Tom Howard

First published in Great Britain in 2025 by
Dorling Kindersley Limited, in association with the Royal Horticultural Society,
DK, One Embassy Gardens, 8 Viaduct Gardens,
London SW11 7BW

The authorised representative in the EEA is
Dorling Kindersley Verlag GmbH. Arnulfstr. 124,
80636 Munich, Germany

Copyright © 2025 Dorling Kindersley Limited
A Penguin Random House Company
10 9 8 7 6 5 4 3 2 1
001–339188–Jan/2025
Text copyright © Lucy Chamberlain 2025
Lucy Chamberlain has asserted her right to be identified as the author of this work

All rights reserved.
No part of this publication may be reproduced, stored
in or introduced into a retrieval system, or transmitted,
in any form, or by any means (electronic, mechanical,
photocopying, recording, or otherwise), without the prior
written permission of the copyright owner.

A CIP catalogue record for this book
is available from the British Library.
ISBN: 978-0-2416-5649-5

Printed and bound in China
www.dk.com

This book was made with Forest Stewardship Council™ certified paper – one small step in DK's commitment to a sustainable future.
Learn more at **www.dk.com/uk/information/sustainability**

The Royal Horticultural Society is the UK's leading gardening charity dedicated to advancing horticulture and promoting good gardening. Its charitable work includes providing expert advice and information in print, online, and at its five major gardens and annual shows, training gardeners of every age, creating hands-on opportunities for children to grow plants, and sharing research into plants, wildlife, wellbeing, and environmental issues affecting gardeners. For more information visit: www.rhs.org.uk or call 020 3176 5800.